AS/A-LEVEL YEAR 1

STUDENT GUIDE

EDEXCEL

Biology B

Topics 3 and 4

Classification and biodiversity

Exchange and transport

Mary Jones

PHILIP ALLAN FOR
HODDER
EDUCATION
AN HACHETTE UK COMPANY

Philip Allan, an imprint of Hodder Education, an Hachette UK company, Blenheim Court, George Street, Banbury, Oxfordshire OX16 5BH

Orders

Bookpoint Ltd, 130 Milton Park, Abingdon, Oxfordshire OX14 4SB

tel: 01235 827827

fax: 01235 400401

e-mail: education@bookpoint.co.uk

Lines are open 9.00 a.m.–5.00 p.m., Monday to Saturday, with a 24-hour message answering service. You can also order through the Hodder Education website: www.hoddereducation.co.uk

© Mary Jones 2015

ISBN 978-1-4718-4387-7

First printed 2015

Impression number 5 4 3 2

Year 2018 2017

This guide has been written specifically to support students preparing for the Edexcel AS and A-level Biology B examinations. The content has been neither approved nor endorsed by Edexcel and remains the sole responsibility of the author.

Cover photo: Elena Pankova/Fotolia

Typeset by Greenhill Wood Studios

Printed in Italy by Printer Trento S.r.l.

Hachette UK's policy is to use papers that are natural, renewable and recyclable products and made from wood grown in sustainable forests. The logging and manufacturing processes are expected to conform to the environmental regulations of the country of origin.

Contents

Content Guidance

Questions & Answers

▮Getting the most from this book

Exam tips

Advice on key points in the text to help you learn and recall content, avoid pitfalls, and polish your exam technique in order to boost your grade.

Knowledge check

Rapid-fire questions throughout the Content Guidance section to check your understanding.

Knowledge check answers

1 Turn to the back of the book for the Knowledge check answers.

Summaries

■ Each core topic is rounded off by a bullet-list summary for quick-check reference of what you need to know.

Exam-style questions ——▶

Commentary on the questions

Tips on what you need to do to gain full marks, indicated by the icon **e**.

Sample student answers

Practise the questions, then look at the student answers that follow.

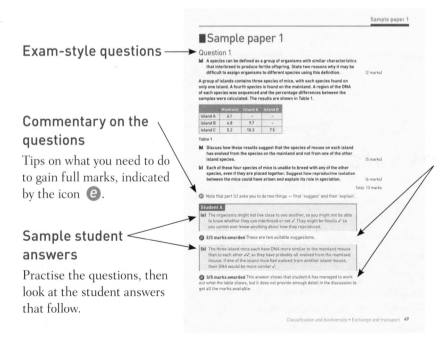

Commentary on sample student answers

Find out how many marks each answer would be awarded in the exam and then read the comments (preceded by the icon **e**) following each student answer. These indicate exactly how and where marks are gained or lost.

■ About this book

This book is the second in a series of four covering the Edexcel AS and A-level Biology B specifications. It covers Topics 3 and 4:

- Classification and biodiversity
- Exchange and transport

This guide has two main sections:

- The **Content Guidance** provides a summary of the facts and concepts that you need to know for these two topics.
- The **Questions & Answers** section contains two specimen papers for you to try, each worth 80 marks. There are also two sets of answers for each question, one from a candidate who is likely to get a C grade and another from a candidate who is likely to get an A grade.

The specification

It is a good idea to have your own copy of the Edexcel Biology B specification. It is you who is going to take this examination, not your teacher, and so it is your responsibility to make sure you know as much about the exam as possible. You can download a copy free from www.edexcel.com.

The AS examination is made up of two papers:

- **Paper 1** Core Cellular Biology and Microbiology (1 hour 30 minutes, 80 marks)
- **Paper 2** Core Physiology and Ecology (1 hour 30 minutes, 80 marks)

The A-level examination is made up of three papers:

- **Paper 1** Advanced Biochemistry, Microbiology and Genetics (1 hour 45 minutes, 90 marks)
- **Paper 2** Advanced Physiology, Evolution and Ecology (1 hour 45 minutes, 90 marks)
- **Paper 3** General and Practical Principles in Biology (2 hours 30 minutes, 120 marks)

This book covers content that will be examined in Paper 2 of the AS examination and Papers 1, 2 and 3 of the A-level examination.

What is assessed?

It is easy to forget that your examination is not just testing what you *know* about biology — it is also testing your *skills*. It is difficult to overemphasise how important these are.

The Edexcel examination tests three different assessment objectives (AOs). The following table gives a breakdown of the proportion of marks awarded to each assessment objective in the AS and A-level examinations.

Assessment objective	Outline of what is tested	Percentage of marks (AS)	Percentage of marks (A-level)
AO1	Demonstrate knowledge and understanding of scientific ideas, processes, techniques and procedures	35–37	31–33
AO2	Apply knowledge and understanding of scientific ideas, processes, techniques and procedures: ■ in a theoretical context ■ in a practical context ■ when handling qualitative data ■ when handling quantitative data	41–43	41–43
AO3	Analyse, interpret and evaluate scientific information, ideas and evidence, including in relation to issues, to: ■ make judgements and reach conclusions ■ develop and refine practical design and procedures	20–23	25–27

AO1 is about remembering and understanding all the biological facts and concepts you have covered. AO2 is about being able to *use* these facts and concepts in new situations. The examination paper will include questions that contain unfamiliar contexts or sets of data, which you will need to interpret in the light of the biological knowledge you have. When you are revising, it is important that you try to develop your ability to do this, as well as just learning the facts.

AO3 is about practical and experimental biology. A science subject such as biology is not just a body of knowledge. Our knowledge and understanding of biology continues to develop as scientists find out new information through their research. Sometimes, new research means that we have to change our ideas.

You will need to develop your skills at doing experiments to test hypotheses and analyse the results to determine whether the hypothesis is supported or disproved. You need to appreciate why science does not always give us clear answers to the questions we ask. You will be asked to make judgements and reach conclusions, and be able to design and improve experiments and procedures whose results we can trust.

Scientific language

Throughout your biology course, and especially in your examination, it is important to use clear and correct biological language. Scientists take great care to use language precisely. If doctors or researchers do not use exactly the correct word when communicating with someone, what they are saying could be easily misinterpreted.

Biology has a huge number of specialist terms and it is important that you learn them and use them. Your everyday conversational language, or what you read in the newspaper or hear on the radio, is often not the kind of language required in a biology examination. Be precise and careful in what you write, so that an examiner cannot possibly misunderstand you.

The examination

Time

In all of the examinations, the mark allocation works out at around 1 minute per mark. When you are trying out a test question, time yourself. Are you working too fast? Or are you taking too long? Get used to what it feels like to work at around a-mark-a-minute rate.

It is not a bad idea to spend one of those minutes just skimming through the exam paper before you start writing. Maybe one of the questions looks as though it is going to need a bit more of your time than the others. If so, make sure you leave a little bit of extra time for it.

Read the question carefully

This sounds obvious, but candidates lose large numbers of marks by not doing it.

- There is often vital information at the start of the question that you will need in order to answer the questions themselves. Do not just jump straight to the first place where there are answer lines and start writing — start reading at the beginning. Examiners are usually careful not to give you unnecessary information, so if it is there it is probably needed. You may like to use a highlighter to pick out any particularly important bits of information in the question.
- Look carefully at the command words (the ones right at the start of the question) and do what they say. For example, if you are asked to *explain* something, you will not get many marks — perhaps none at all — if you *describe* it instead. You can find all these words in Appendix 7 at the end of the A-level specification document.

Depth and length of answer

Examiners will give you two useful guidelines about how much you need to write:

- **The number of marks**. Obviously, the more marks, the more information you need to give. If there are 2 marks, then you will need to give two different pieces of information in order to get both of them. If there are 5 marks, you will need to write much more.
- **The number of lines**. This is not such a useful guideline as the number of marks, but it can still help you to know how much to write. If you find your answer will not fit on the lines, you probably have not focused sharply enough on the question. The best answers are short and precise.

Mathematical skills

Like all of the sciences, biology uses mathematics extensively. The specification contains an appendix that lists and describes the mathematical techniques that you need to be familiar with. You will probably have met most of these before, but make sure that you are confident with all of them. If there are any of which you are uncertain, do your best to improve your skills in them early on the course — do not leave it until the last minute, just before the exam. The more you practise your maths skills, the more relaxed you will be about them in the exam.

Content Guidance

■ Topic 3 Classification and biodiversity

Classification

Biologists classify organisms according to how closely they believe they are related to one another. Each species has evolved from a previously existing species. We do not usually have any information about these ancestral species, so we judge the degree of relatedness between two organisms by looking carefully at their physiology, anatomy and biochemistry. The greater the similarities, the more closely they are thought to be related.

Classification systems

The system used for classification is a **taxonomic system**. This involves placing organisms in a series of taxonomic units to form a **hierarchy**. The largest unit is the domain. Domains are subdivided into kingdoms, phyla, classes, orders, families, genera and species, as shown in Figure 1.

Domain	Eukarya
Kingdom	Animalia
Phylum	Chordata
Class	Mammalia
Order	Rodentia
Family	Muridae
Genus	*Mus*
Species	*Mus musculus* (house mouse)

Figure 1 An example of classification

Knowledge check 1

A king cobra, *Ophiophagus hannah*, belongs to the order Squamata and the family Elapidae. Show the full classification for a king cobra.

Until quite recently, the highest level of classification was the kingdom. Biologists classified all living organisms into the following five kingdoms:

■ **Prokaryota** — these are organisms with prokaryotic cells. This kingdom includes bacteria and blue-green algae.
■ **Protoctista** — these organisms have eukaryotic cells. They mostly exist as single cells, but some are made of groups of similar cells.
■ **Fungi** — fungi have eukaryotic cells surrounded by a cell wall, but this is not made of cellulose and fungi never have chloroplasts.
■ **Plantae** — these are the plants. They have eukaryotic cells surrounded by cellulose cell walls and they feed by photosynthesis.
■ **Animalia** — these are the animals. They have eukaryotic cells with no cell wall.

However, in the 1990s new information led to the proposal of a new classification. This resulted from new discoveries about the molecular biology and metabolic pathways (sequences of chemical reactions) taking place in prokaryotic organisms. Although all prokaryotic organisms have structural similarities, it was discovered that they can have very different molecular biology and biochemistry. These include differences in:

- the structure of their cell membranes and cell walls
- the structure of their flagella (whip-like extensions that can move the cell through water)
- their ribosomal RNA
- the way in which information in DNA is used to build proteins

As a result of this new information, the prokaryotes are now classified as two **domains**, the **Archaea** and the **Bacteria**. They are thought to be as widely different from each other as humans are from bacteria. All other organisms are classified in the domain **Eukarya**. Figure 2 shows the two systems of classification.

Five kingdoms

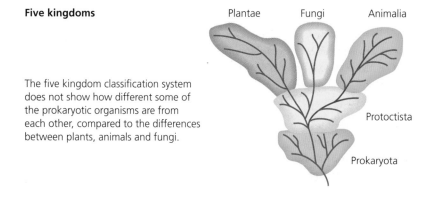

The five kingdom classification system does not show how different some of the prokaryotic organisms are from each other, compared to the differences between plants, animals and fungi.

Three domains

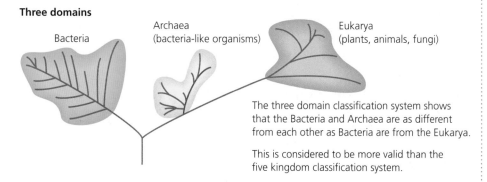

The three domain classification system shows that the Bacteria and Archaea are as different from each other as Bacteria are from the Eukarya.

This is considered to be more valid than the five kingdom classification system.

Figure 2 Two classification systems

This classification is now widely accepted in the scientific community, following the sharing of results from research and peer review (see 'Validating new evidence', pages 12–13). However, there are still a few well-respected scientists who disagree with the three-domain system and researchers continue to collect evidence to test the relationships between organisms belonging to these groups.

Species

A commonly used definition of the term species is 'A group of organisms with similar characteristics that normally interbreed to produce fertile offspring'. However, it is often difficult to apply this definition because:

- two groups of organisms may live in different parts of the world and so we cannot know whether or not they would 'normally' interbreed if they got the chance
- a group of organisms may reproduce only asexually, not sexually, so they can never interbreed
- organisms may be known only from museum specimens and nothing is known about their behaviour in the wild
- extinct organisms exist only as fossils

Knowledge check 2

Suggest why it is particularly difficult to decide if two fossils belong to the same or different species.

Comparing DNA or protein composition

Gel electrophoresis is an important tool in determining the degree of differences between species in order to try to decide how closely related they are. It is also used to decide whether or not two populations belong to the same species or to different species.

Electrophoresis is a way of separating macromolecules of different sizes or bearing different electrical charges. For example, it can be used to separate polypeptides or lengths of DNA of different sizes by applying an electrical potential difference across a gel in which a sample is placed. Lengths of DNA or polypeptides of different masses, or with different charges, will move across the gel at different speeds.

Samples of DNA from the same regions of the chromosomes are taken from individuals of both of the populations to be compared. Each sample of DNA is exposed to a set of **restriction enzymes (restriction endonucleases)**. These enzymes cut DNA molecules where particular base sequences are present. For example, a restriction enzyme called *Bam*H1 cuts where the base sequence GGATCC is present on one strand of the DNA. Other restriction enzymes target different base sequences. This cuts the DNA into fragments of different lengths.

To carry out gel electrophoresis, a small, shallow tank is partly filled with a layer of agarose gel. A potential difference is applied across the gel so that a direct current flows through it.

A mix of the DNA fragments to be separated is placed in wells on the gel. DNA fragments carry a small negative charge, so they slowly move towards the positive terminal. The larger they are, the more slowly they move. After some time, the current is switched off and the DNA fragments stop moving through the gel.

The DNA fragments must be made visible in some way so that their final positions can be determined. This can be done by adding fluorescent markers to the fragments. Alternatively, single strands of DNA made using radioactive isotopes and with base sequences thought to be similar to those in the DNA fragments can be added to the gel. These are called **probes**. They will pair up with fragments that have complementary base sequences, so their positions are now emitting radiation. This can be detected by its effects on a photographic plate.

By comparing the patterns obtained by analysing DNA from the same locus in two or more groups of organisms, we can determine how similar their DNA base sequences are and therefore how closely related they are, as shown in Figure 3.

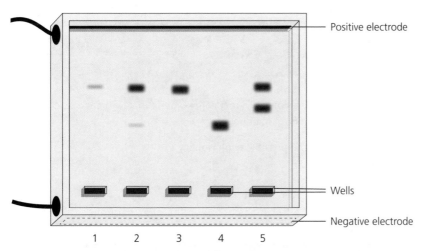

Figure 3 Electrophoresis gel comparing the enzyme glucose-6-phosphate dehydrogenase in three species of the parasitic protoctistan *Leishmania*

Electrophoresis of polypeptides is done in a similar way, but of course without having to cut them up using restriction enzymes first. A protein that is often used for comparison is cytochrome b, found in mitochondria. Differences in the amino acid sequences in the polypeptide may cause differences in its overall charge and therefore the rate at which it moves across the gel. There may also be differences in mass, resulting from different numbers of amino acids in the polypeptide or amino acids with different R groups.

Exam tip

Always make it clear whether you are writing about comparing proteins or DNA — don't confuse the two.

Knowledge check 3

a What do the bars in Figure 3 represent?
b What causes the bars to appear at different distances from the cathode on the gel?
c What evidence is there in Figure 3 that the samples of the enzyme were taken from three different species?

DNA sequencing

DNA sequencing involves determining the exact sequence of bases in a length of DNA. If two groups of organisms have similar base sequences in comparable regions of their DNA, we can say that they belong to the same species. If there is always the same set of differences between them, we may decide that they belong to different species (so long as we have no indication that they can interbreed to produce fertile offspring). The more similar their DNA, the more recently the species are considered to have diverged from one another, as shown in Figure 4.

```
              1        10        20        30        40        50        60
Human        ATGGGTGATGTTGAGAAAGGCAAGAAGATTTTTATTATGAAGTGTTCCCAGTGCCACACC
Chimpanzee   ..........................................................T....
Rhesus monkey .........................................G...
Mouse        ..........A.............G..CA......G.............T
Chicken      .....A...A......G......G.CCA...A......T..G
Clawed toad  .....A..............AG.C...G.CCA...A......T
Rainbow trout .........A...CT..G..A......GCA...G.CCA......G.......T..T..T
```

Figure 4 Differences in the base sequence of the first 60 bases of the cytochrome c gene in seven species of vertebrates. A dot means that the base is the same as in the human gene.

By analysing differences in DNA sequences, we can trace the possible evolutionary relationships and evolutionary histories of different species.

Knowledge check 4

a Copy and complete the table to show the number of differences between the base sequences (Figure 4) for the seven species of vertebrates.

	Human	Chimpanzee	Rhesus monkey	Mouse	Chicken	Clawed toad
Chimpanzee	1					
Rhesus monkey	1	2				
Mouse						
Chicken						
Clawed toad						
Rainbow trout						

b Which species appear to have shared a common ancestor most recently?
c Which two species appear to be most distantly related?

Bioinformatics

Bioinformatics is the use of large databases and computer software to store and analyse information about living organisms. For example, there are now databases listing the base sequences of all the DNA in several individual people, as well as databases for the genomes of other species such as the nematode worm *Caenorhabditis* and the malarial parasite *Plasmodium*. There are also databases for the proteomes of different species. (The proteome is all the different proteins in an organism.)

There are many different uses that can be made of such databases, including research into the closeness of the relationships between species.

Validating new evidence

As it has become easier and cheaper to analyse DNA and proteins in different individuals, populations and species, researchers are finding evidence for relationships between species that differ from those previously believed to be correct. As with all results from research, it is important that this evidence is validated before it can be widely accepted. The standard process that allows this to happen includes the following steps.

- Researchers submit papers, which include details of their methods, results and conclusions, to a scientific journal.
- The journal sends the paper to another expert in the field to be checked carefully. This is called peer review.
- If the paper is judged to be sound, it is published in the journal.
- Other groups of researchers attempt to repeat the methods, to check if they get the same results.

Face-to-face discussions are also an important part of the review process. Conferences take place on particular subject areas, which are attended by researchers in that field from all over the world. This provides an opportunity for scientists to discuss on-going research that has not yet been published, to share ideas and to challenge each other's ideas.

Summary

After studying this topic, you should be able to:
- list the hierarchy of groups into which living organisms are classified
- explain why the three-domain system has been introduced
- state a definition of a species and explain why it is sometimes difficult to apply this definition
- explain how gel electrophoresis can be used to identify differences between sections of DNA or polypeptides in different populations and therefore provide evidence about possible evolutionary relationships between species
- state that DNA sequencing and bioinformatics can be used to help determine evolutionary relationships and distinguish between species
- explain how new evidence is validated by the scientific community

Natural selection

Adaptations

Each species of organism has its own particular role to play within a community of living organisms. This is called its **niche**. It includes the ways in which it interacts with other organisms — for example, what it eats and what eats it — and also the ways in which it interacts with its environment — for example, the level of humidity it requires for survival or the kind of nesting site it needs.

For example, dandelions require light in order to be able to photosynthesise and are adapted for growing where light is fairly bright. They require moist soil but cannot grow where the soil is waterlogged. They do not thrive in shaded woodland conditions. Dandelion leaves and flowers are eaten by many different herbivores and they are able to grow close to the ground so that few herbivores are able to eat absolutely all of their leaves. Their fruits (dandelion clocks) contain seeds that are dispersed by wind. Each species of organism has a set of **adaptations** that allow it to live and reproduce in a particular environment. These adaptations may be behavioural, physiological and anatomical (Table 1).

Exam tip

Don't confuse the term 'niche' with 'habitat'. An organism's niche is not a place; it is the role that it has in an ecosystem.

Organism	Behavioural adaptation	Physiological adaptation	Anatomical adaptation
Earthworm	Responds extremely rapidly to a bird peck on its head by retracting into its burrow. Generally only comes out to feed at night	Has blood containing haemoglobin that absorbs oxygen even in the relatively low oxygen concentrations underground	Has stiff hairs called chaetae along its lower surface that grip firmly on to the sides of its burrow and help it to move underground
Venus fly trap	Leaves fold when sensitive hairs are touched	Cells in leaf surface secrete hydrolytic enzymes that digest trapped insects	Leaves have a fringe of still hairs that prevents insects escaping when the leaf has folded
Salmon	When adult, leaves the sea and swims upstream along the river in which it was born to find a place to spawn	Gills are able to switch from excreting salt when in the sea to taking it up by active transport when in fresh water	Adult fish have unusually strong swimming muscles that allow them to swim upstream in fast-flowing rivers and even to leap up waterfalls

Table 1 Some examples of adaptations

Natural selection

In any population of organisms, genes come in various forms known as **alleles**. **Mutation** may produce new alleles. Different combinations of alleles in different individuals produce variation in their characteristics. Individuals with particular characteristics may be better adapted to their environment and may therefore have an increased chance of surviving long enough to reproduce. This process is called **natural selection**.

Alleles that produce these advantageous characteristics are therefore more likely to be passed on to the next generation. Over time and many generations, these alleles will become more common in the population while other, less advantageous, alleles become less common. **Evolution** happens when there is a change in allele frequency over time.

Usually, a species is already well adapted to its environment, so allele frequency remains fairly constant over many generations. However, if the environment changes (for example, if the climate becomes warmer) or if a new advantageous allele arises by mutation, natural selection may produce a change in allele frequency.

Drug resistance in bacterial pathogens

Antibiotics are drugs that are given to people who have infections caused by bacteria. Antibiotics kill bacteria but do not normally harm human cells.

In a population of bacteria, some may by chance have an allele that confers resistance to a particular antibiotic. These resistance genes are often present in plasmids. The widespread use of antibiotics to destroy bacterial pathogens provides a selection pressure that favours the survival of individual bacteria that possess alleles that provide resistance. These bacteria survive and reproduce while non-resistant bacteria are killed. This results in a whole population of bacteria that are not killed by the antibiotic.

Exam tip

When describing adaptations, try to avoid any wording that would imply that the organism has purposefully developed that adaptation. Adaptations arise as a result of natural selection, as described below.

Knowledge check 5

Construct your own table to describe the adaptations of two other organisms.

Exam tip

Take care to use the terms 'allele' and 'gene' correctly. Alleles are forms of a gene.

Pharmaceutical companies therefore need to try to develop new antibiotics against which bacteria have not (as yet) evolved resistance. This can be seen as an 'evolutionary race' as bacteria evolve resistance against new drugs.

Speciation

Speciation is the production of new species. One pre-existing species becomes two separate species. The crucial event that must occur for this to happen is that one population must become unable to interbreed with another. They must become **reproductively isolated** from one another. Once this has happened, we can say that the two populations are now different species.

Allopatric speciation

There are many ways in which reproductive isolation can happen. One that we think has been especially important in the formation of new species of plants and animals begins by a group of individuals in the population becoming **geographically isolated** from the rest. This is called **allopatric speciation**.

For example, a few lizards might get carried out to sea on a floating log and be transported to an island where that species of lizard was not previously found. This island group is subjected to different environmental conditions from the rest of the species left behind on the mainland. Different alleles are therefore selected for in the two groups. Over time, the allele frequency in the island group becomes very different from the allele frequency in the original, mainland lizards. This may cause their characteristics to become so different that — even if a bridge appears between the two islands — they can no longer interbreed to produce fertile offspring. Reasons for this could include:

- they have evolved different courtship behaviours so that mating no longer occurs between them
- the sperm of one group are no longer able to survive in the bodies of the females of the other group, so fertilisation does not occur
- the number or structure of the chromosomes is different, so that the zygote that is formed by fertilisation does not have a complete double set of genes and cannot develop
- even if a zygote is successfully produced, the resulting offspring may not be able to form gametes because its two sets of chromosomes (one from each parent) are unable to pair up with each other successfully and so cannot complete meiosis

Sympatric speciation

Speciation can also occur when the two populations are living in the same area. This is called **sympatric speciation**. For example:

- During meiosis to form gametes, homologous chromosomes may not separate properly, producing a gamete with two sets of chromosomes instead of one. When this fuses with a normal haploid gamete, a triploid zygote is formed. This is called **polyploidy**. Polyploidy has led to the production of many new plant species. The triploid plant is able to reproduce asexually, but it cannot produce viable gametes because the odd number of each kind of chromosome (three) means that they cannot pair up correctly during meiosis. Therefore, it cannot reproduce with members of the original species.

- Different behaviours may develop in two populations of animals — for example, one population may learn to hunt mostly at night while another population hunts during the daytime. Individuals from each group rarely meet each other, so they develop different adaptations and do not have the opportunity to mate and exchange genes with one another. Just as in allopatric speciation, this can eventually lead to behavioural, physiological and anatomical differences, preventing interbreeding.

Summary

After studying this topic, you should be able to:
- explain what is meant by the terms 'niche' and 'adaptation'
- understand how organisms occupy niches according to their behavioural, physiological and anatomical adaptations
- explain how evolution can come about through natural selection acting on variation, bringing about adaptations
- explain why there is an 'evolutionary race' between bacteria and the development of new antibiotics
- explain what is meant by the term 'speciation'
- explain the importance of reproductive isolation in the process of speciation
- describe how allopatric and sympatric speciation may occur

Biodiversity

Biodiversity includes:
- the range of ecosystems or habitats
- the numbers of different species, and
- the genetic variation that exists within the populations of each species in a particular area.

Assessing biodiversity

Calculating an index of diversity

We can find out something about the biodiversity of an area by measuring the **distribution** and **abundance** of the different species present. You cannot usually count every single organism in the area, so a **sampling** technique is used. It is important that this is random in order to avoid bias in your choice of area in which to make your measurements. For example, you could use a mobile phone app to generate random numbers. These numbers can be used as x and y coordinates to tell you where to place a **quadrat** within a defined area. You then count up the numbers of each species in the quadrat. Repeat with at least nine more quadrats.

A habitat that has more species than another is said to have greater species richness. However, in order to compare the biodiversity of each area, we also need to take into account the relative numbers of each species. For example, if area A has ten species, with almost all of the organisms belonging to one or two species, it has a lower biodiversity than area B, which also has ten species but with the numbers of organisms spread evenly between them.

Exam tip

The correct spelling is quadrat, not quadrant.

One method of comparing the biodiversity of different areas is to use an index of diversity. This can be calculated using the formula:

$$D = \frac{N(N-1)}{\Sigma n(n-1)}$$

where D is the index of diversity, N is the total number of individuals of all species in your sample, n is the total number of individuals of one species and Σ is the sum of.

The greater the value of D, the greater the diversity.

Knowledge check 6

Use this formula to calculate an index of diversity of an area from which these results were obtained.

Species	Number of individuals, n
A	2
B	35
C	1
D	81
E	63
F	2
G	5
H	11
I	1
	Total number of individuals = 201

Exam tip

Remember always to show every step in your working or you may not be given full credit, even if your final answer is correct.

Assessing genetic diversity

Biodiversity can also be assessed within a species by measuring the variety of alleles that are present in the gene pool of a population. This involves finding out how many genes have different alleles and how many alleles each of them has. This is best done using DNA sequencing.

The importance of biodiversity

Maintaining biodiversity is important for many reasons.

Stability of ecosystems

- The loss of one of more species within a community may have negative effects on others so that eventually an entire ecosystem becomes seriously depleted.
- Genetic diversity within a species increases the likelihood that the species will be able to evolve to survive in a changing environment, such as following the introduction of a new species or as a result of climate change.

Benefits to humans

- The loss of a species could prevent an ecosystem from being able to supply humans with their needs. For example, an ancient civilisation in Peru, the Nazca people, are thought to have cut down so many huarango trees that their environment became too dry and barren for them to grow crops.

■ We make use of many different plants to supply us with drugs. There may be many more species of plants, for example in rainforests, which could potentially provide life-saving medicines.

■ In some countries such as Kenya, people derive income from tourists who visit the country to see wildlife.

■ Several scientific studies have found that people are happier and healthier if they live in an environment with high biodiversity.

Moral and ethical reasons

■ Many people feel that it is clearly wrong to cause the extinction of a species and that we have a responsibility to maintain environments in which all the different species on earth can live.

Conservation

Conservation aims to retain and possibly to increase biodiversity. Conservation is a dynamic process involving active management. Conservation can be carried out:

■ in situ — in the natural environment where organisms live
■ ex situ — in places such as zoos and botanic gardens

Habitat conservation

The best way to conserve threatened species is in their own natural habitat. For example, National Parks, marine parks and nature reserves can set aside areas of land or sea in which species are protected. A species will only survive if the features of its habitat that are essential to its way of life are maintained. Maintaining habitat for one species is also likely to be beneficial to many other species in the community.

However, this is often difficult because people living in that area have their own needs. In parts of the world where people are already struggling to survive, it is difficult to impose restrictions on the way they can use the land where they live. The best conservation programmes involve the local people in habitat conservation and reward them for it in some way. For example, they could be given employment in a National Park or be paid for keeping a forest in good condition.

A habitat that has been severely damaged is said to be **degraded**. It is sometimes possible to restore degraded habitats. For example, a heavily polluted river or one that has been over-engineered (i.e. its channel has been straightened or its banks have been concreted over) can be restored to a more natural state.

Zoos

Zoos often run **captive breeding programmes**. This involves collecting together a small group of organisms of a threatened species and encouraging them to breed together. In this way, extinction can be prevented. The breeding programme will try to maintain or even increase genetic diversity in the population by breeding unrelated animals together. This can be done by moving males from one zoo to another or by *in vitro* fertilisation using frozen sperm transported from males in another zoo. It may also be possible to implant embryos into a surrogate mother of a different species so that many young can be produced, even if there is only a small number of females of the endangered species. Eggs and embryos, like sperm, can be frozen and stored for long periods of time. Collections of frozen sperm, eggs or embryos are sometimes known as 'frozen zoos'.

The best captive breeding programmes work towards the **reintroduction** of individuals to their original habitat, if this can be made safe for them. It is important that work is done on the ground to prepare the habitat for the eventual reintroduction of the animals. For example, the scimitar-horned oryx has been successfully reintroduced to Tunisia. This followed a widespread captive breeding programme in European zoos and the preparation and protection of suitable habitat, including the education and involvement of people living in or around the proposed reintroduction area. Captive breeding followed by reintroduction is an example of the use of both in-situ and ex-situ conservation.

Zoos can use **education** programmes to bring conservation issues to the attention of large numbers of people who may decide to contribute financially towards conservation efforts or to campaign for them. Entrance fees and donations can be used to fund conservation programmes both in the zoo itself and in natural habitats.

Research into animals in zoos can find out more about their needs in terms of food, breeding places etc. This can help to inform people working on conservation in natural habitats.

Botanic gardens

Botanic gardens are similar to zoos but they are for plants rather than animals. Like zoos, they can be safe havens for threatened plant species and are involved in breeding programmes, reintroduction programmes, education and research.

Seed banks

Seed banks store seeds collected from plants. Many seeds live for a long time in dry conditions, but others need more specialised storage environments. A few of the seeds are germinated every so often so that fresh seed can be collected and stored.

Seed banks can help the conservation of plants just as zoos can help the conservation of animals. The Royal Botanic Gardens at Kew has a huge seed bank based at Wakehurst Place, Sussex, UK. Collectors search for seeds, especially those of rare or threatened species, and bring them to the seed bank where they are stored carefully. Another seed bank, built into the permafrost (permanently frozen ground) in Norway, aims to preserve seeds from all the world's food crops.

Summary

After studying this topic, you should be able to:
- explain what is meant by the term 'biodiversity'
- calculate an index of diversity to assess diversity at the species level in a habitat
- describe genetic diversity in terms of the variety of alleles in the gene pool of a population
- explain the biological, ethical and economic reasons for maintaining biodiversity
- explain the principles of in-situ (protected habitats) and ex-situ (zoos and seed banks) conservation and discuss issues relating to each type of conservation

■ Topic 4 Exchange and transport

Surface area to volume ratio

Living organisms constantly exchange substances with their environment. For example, organisms whose cells respire aerobically take in oxygen from the air or water that surrounds them and release carbon dioxide. These gases move into and out of the organisms through its surface, by diffusion.

The rate at which the gases can diffuse into and out of the organism depends on the surface area of the exchange surface. The rate at which the gases are used is affected by the volume of the organism.

In a small organism such as the unicellular organism *Amoeba*, the ratio of surface area to volume is relatively large. Diffusion across the cell surface is fast enough to supply all of the cell's needs.

In a large organism such as an insect, a fish or a mammal, the surface area to volume ratio is less. Diffusion across the general body surface would not allow enough oxygen to enter fast enough for the needs of the respiring cells. Moreover, diffusion from the exchange surface to all the body cells would take far too long. Larger organisms therefore tend to have:

■ specialised exchange surfaces that provide a large surface area across which exchange of substances with the environment can take place
■ a mass transport system that carries substances between the exchange surfaces and body cells

Summary

After studying this topic, you should be able to:
■ explain how surface area and volume affect the exchange of substances between an organism and its environment
■ state that surface area to volume ratio decreases as size increases
■ explain why large organisms need specialised exchange surfaces and mass transport systems

Cell transport mechanisms

Substances enter and leave cells by moving across the cell surface membrane.

Structure of a cell surface membrane

A cell surface membrane consists of a double layer of **phospholipid** molecules. This structure arises because in water a group of phospholipid molecules arranges itself into a **bilayer**, with the hydrophilic heads facing outwards into the water and the hydrophobic tails facing inwards, therefore avoiding contact with water. This is the basic structure of a cell membrane (Figure 5).

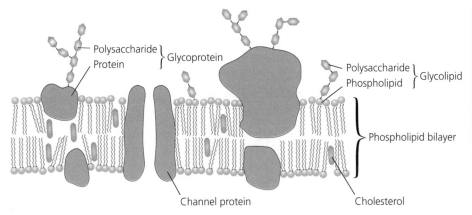

Figure 5 The fluid mosaic model of membrane structure

There are also **cholesterol** molecules among the phospholipids. **Protein** molecules float in the phospholipid bilayer. Many of the phospholipids and proteins have short chains of carbohydrates attached to them, on the outer surface of the membrane. They are known as **glycolipids** and **glycoproteins**. There are also other types of glycolipid with no phosphate groups. The roles of these components are outlined in Table 2.

Component	Roles
Phospholipids	Form the fluid bilayer that is the fundamental structure of the membrane Prevent hydrophilic substances such as ions and some molecules from passing through
Cholesterol	Helps to keep the cell membrane fluid
Proteins and glycoproteins	Provide channels that allow hydrophilic substances to pass through the membrane. These channels can be opened or closed to control the substances' movement Actively transport substances through the membrane against their concentration gradient using energy derived from ATP Act as receptor molecules for substances such as hormones, which bind with them. This can then affect the activity of the cell Cell recognition — cells from a particular individual or a particular tissue have their own set of proteins and glycoproteins on their outer surfaces
Glycolipids	Cell recognition and adhesion to neighbouring cells to form tissues

Table 2 The roles of the components of cell membranes

This is called the fluid mosaic model of membrane structure:

- 'fluid' because the molecules within the membrane can move around within their own layers
- 'mosaic' because the protein molecules are dotted around within the membrane
- 'model' because no one has ever seen a membrane looking like the diagram — the molecules are too small to see even with the most powerful microscope. The structure has been worked out because it explains the behaviour of membranes that has been discovered through experiment

Passive transport through cell membranes

Molecules and ions are in constant motion. In gases and liquids they move freely. As a result of their random motion, each type of molecule or ion tends to spread out evenly within the space available. This is diffusion. Diffusion results in the net movement of ions and molecules from a high concentration to a low concentration.

Diffusion across a cell membrane

Some molecules and ions are able to pass through cell membranes. The membrane is permeable to these substances. However, some substances cannot pass through cell membranes, so the membranes are said to be **partially permeable**.

For example, oxygen is often at a higher concentration outside a cell than inside it because the oxygen inside the cell is being used up in respiration. The random motion of oxygen molecules inside and outside the cell means that some of them 'hit' the cell membrane. More of them hit the membrane on the outside than the inside because there are more of them outside. Oxygen molecules are small and do not carry an electrical charge, so they are able to pass freely through the phospholipid bilayer. Oxygen therefore diffuses from outside the cell, through the membrane, to inside the cell, down its concentration gradient. This is known as passive transport because the cell does not do anything to cause the oxygen to move across the cell membrane.

Facilitated diffusion

Ions or electrically charged molecules are not able to diffuse through the phospholipid bilayer because they are repelled from the hydrophobic tails. Large molecules are also unable to move freely through the phospholipid bilayer. However, the cell membrane contains special protein molecules that provide hydrophilic passageways through which these ions and molecules can pass. The proteins that form these passageways are called **channel proteins** and **carrier proteins**. Channel proteins form pores that can usually be opened or closed like gates; they are called gated channels. Carrier proteins, on the other hand, change shape to allow molecules or ions to bind with them on one side of the membrane and then release the molecules or ions on the other side. Different channel and carrier proteins allow the passage of different types of molecules and ions. Diffusion through these channel and carrier proteins is called **facilitated diffusion**. Like 'ordinary' diffusion, it is entirely passive. It always takes place down a concentration gradient and does not require any input from the cell.

Figure 6 shows both diffusion and facilitated diffusion across a cell surface membrane.

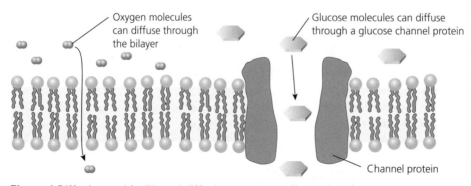

Oxygen molecules can diffuse through the bilayer

Glucose molecules can diffuse through a glucose channel protein

Channel protein

Figure 6 Diffusion and facilitated diffusion across a cell membrane

Exam tip

Take care not to confuse facilitated diffusion — in which substances move down their concentration gradient — with active transport, in which they move up the gradient.

Knowledge check 7

Explain the similarities and differences between diffusion and facilitated diffusion.

Osmosis

Water molecules are small. They carry tiny electrical charges (dipoles), but their small size means that they are still able to move quite freely through the phospholipid bilayer of most cell membranes. Therefore, water molecules tend to diffuse down their concentration gradient across cell membranes.

Cell membranes always have a watery solution on each side. These solutions may have different concentrations of solutes. The greater the concentration of solute, the less water is present. The water molecules in a concentrated solution are also less free to move because they are attracted to the solute molecules. A concentrated solution is therefore said to have a low water potential. In a dilute solution, there are more water molecules and they can move more freely. This solution has a high water potential.

Imagine a cell membrane with a dilute solution on one side and a concentrated solution on the other side. The solute has molecules that are too large to pass through the membrane — only the water molecules can get through. Water molecules in the dilute solution are moving more freely and therefore hit the membrane more often than water molecules in the concentrated solution. More water molecules therefore diffuse across the membrane from the dilute solution to the concentrated solution than in the other direction. The net movement of water molecules is from a high water potential to a low water potential, down a water potential gradient. This is osmosis (Figure 7).

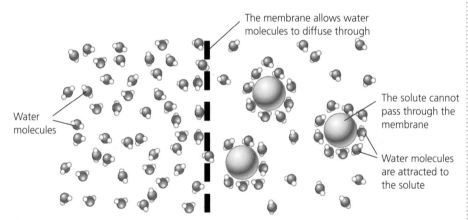

The membrane allows water molecules to diffuse through

Water molecules

The solute cannot pass through the membrane

Water molecules are attracted to the solute

Figure 7 Osmosis across a cell membrane

The water potential of a solution is affected by:

- the concentration of solute in it. The greater the concentration of solute, the lower the water potential. The contribution of the solute to the water potential of the solution is called the **solute potential**.
- the pressure exerted on the solution. The greater the pressure, the higher the water potential. The contribution of pressure to the water potential of the solution is called the **pressure potential**.

  water potential = solute potential + pressure potential

In a plant cell, for example, the water potential is affected by:

- the concentration of the cytoplasm and cell sap

■ the inwards pressure of the cell wall on the cell contents, known as wall pressure. The wall pressure is an equal and opposite force to the outward force of the cell contents on the cell wall, known as **turgor pressure** (Figure 8).

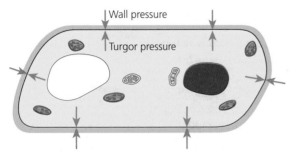

Figure 8 Turgor pressure

Knowledge check 8

If the turgor pressure of a cell increases, will this increase or decrease the water potential of its contents?

Endocytosis and exocytosis

Cells can move substances into and out of the cell without the substances having to pass through the cell membrane. Endocytosis and exocytosis can transport molecules that are too large to pass through the cell membrane.

In **endocytosis**, the cell puts out extensions around the object to be engulfed. The membrane fuses together around the object, forming a vacuole.

In **exocytosis**, the object is surrounded by a membrane inside the cell to form a vacuole or vesicle. This is then moved to the cell membrane. The membrane of the vacuole or vesicle fuses with the cell membrane, expelling its contents outside the cell.

Exocytosis

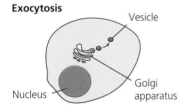

A vesicle is produced containing material to be removed from the cell.

The vesicle moves to the cell surface membrane.

The membrane of the vesicle and the cell surface membrane join and fuse.

The contents of the vesicle are released.

Endocytosis

The cell surface membrane grows out so the object or solution is surrounded.

The membrane breaks and rejoins, enclosing the object.

The vesicle moves inwards and its contents are absorbed.

Figure 9 Endocytosis and exocytosis

Knowledge check 9

Suggest whether endocytosis and exocytosis require the input of energy from the cell. Explain your answer.

Core practical 5

Investigate the effect of temperature on beetroot membrane permeability

Beetroot cells contain a red pigment that cannot pass through the cell membrane. If the structure of the membrane is damaged, the pigment can leak out. By measuring how much pigment leaks out in a particular length of time, the degree of damage to the membrane can be assessed.

Use a cork borer to cut several cylinders of beetroot, preferably all from the same root. Cut each cylinder into several smaller pieces, all exactly the same size. Wash all of them thoroughly in distilled water to remove any red pigment from cells that have been cut open.

Prepare tubes containing water at different temperatures. If possible, use three tubes for each temperature. Keep all other variables constant, such as the volume of the liquid. Immerse pieces of beetroot into each tube. Leave for a standard amount of time.

Remove the beetroot pieces from the tubes. Use a colorimeter with a green filter to measure the absorbance of the liquid in each tube. Calculate the mean absorbance for each temperature. The greater the absorbance, the greater the quantity of beetroot pigment that has leaked out from the cells, indicating a greater degree of damage to the membrane.

High temperatures increase membrane permeability because all the molecules in the membrane move around faster, creating temporary gaps between phospholipid molecules through which other molecules can pass. High temperatures also break hydrogen bonds in protein molecules so channel proteins and carrier proteins lose their shapes, creating gaps through which other molecules can pass.

Core practical 6

Determine the water potential of a plant tissue

Cut cylinders, discs or strips of a solid and uniform plant tissue, such as a potato tuber, then measure either their lengths or masses before immersing them in the solutions. Leave them long enough for them to come to equilibrium and then measure the length or mass of each piece again. Calculate the percentage change in the measurement. Percentage change can then be plotted against the concentration of the solution. The concentration of the solution that gives (or would give) no change in length or mass has the same water potential as the contents of the cells in the tissue.

Knowledge check 10

Draw a graph to predict the results of an experiment to investigate the effect of temperature on membrane permeability. Label each axis fully.

Active transport across cell membranes

Cells are able to make some substances move across their membranes up their concentration gradients. For example, there may be more potassium ions inside the cell than outside the cell. The potassium ions would therefore diffuse out of the cell.

However, the cell may require potassium ions. It may therefore use a process called active transport to move potassium ions from outside the cell to inside the cell, against the direction in which they would naturally diffuse.

This is done using **carrier proteins** in the cell membrane. Each carrier protein is specific to only one type of ion or molecule. Cells contain many different carrier proteins in their membranes. These use energy to move molecules or ions into the cell.

Glucose molecule ⎯⎯⎯
Carrier protein ⎯⎯
Cell membrane ⎯

1 A glucose molecule enters the carrier protein.

2 The carrier protein changes shape. The energy needed to do this comes from ATP.

3 The change of shape of the carrier protein pushes the glucose molecule into the cell.

Figure 10 Active transport

The energy used in active transport comes from the hydrolysis of ATP:

$$ATP \rightarrow ADP + phosphate + energy$$

ATP can be resynthesised by using energy, supplied by respiration, to phosphorylate ADP. ATP provides an immediately and easily accessible supply of energy for a wide range of biological processes.

Knowledge check 11

Explain how active transport differs from facilitated diffusion.

Summary

After studying this topic, you should be able to:
- describe the fluid mosaic model of membrane structure
- explain how the following methods of passive transport take place: diffusion, facilitated diffusion and osmosis
- explain the difference between channel proteins and carrier proteins
- describe how large molecules can be transported out of and into cells by exocytosis and endocytosis
- describe how to investigate the effect of temperature on beetroot membrane permeability
- describe how to determine the water potential of a plant tissue
- describe how active transport takes place
- explain the role of ATP in active transport
- know that the hydrolysis of ATP releases energy for use by the cell and that the production of ATP by the phosphorylation of ADP requires an input of energy

Gas exchange

All organisms exchange gases with their environment. The gases enter and leave their bodies by diffusion, across a gas exchange surface.

Properties of gas exchange surfaces

- They must have a large surface area so that many molecules of gases can move across them at the same time.
- They must be thin so that gases can diffuse across them quickly.

- They must be provided with a good supply of air containing oxygen (or, in an aquatic organism, water containing dissolved oxygen).
- The oxygen must be taken away rapidly from the surface, such as by the flow of blood.

These last two points help to maintain a concentration gradient for oxygen across the gas exchange surface. The greater the concentration gradient, the more rapidly diffusion will occur. This also applies to the concentration gradient for carbon dioxide.

For terrestrial animals, it is also important that the gas exchange surface is not directly exposed to the external air otherwise too much water vapour would be lost. Terrestrial animals tend to have gas exchange surfaces tucked away inside their bodies.

Gas exchange in mammals

In mammals, the alveoli inside the lungs are the gas exchange surface (Figure 11). The adaptations for efficient gas exchange include the following:

- All the alveoli together have a huge surface area because there are many of them and they are very small.
- The wall of an alveolus is only one cell thick and so is the wall of the blood capillary.
- Breathing movements (ventilation) constantly bring fresh air into the lungs and remove air from which oxygen has been lost and to which carbon dioxide has been added in the alveoli.
- Oxygenated blood is taken away from the alveoli in the pulmonary veins.

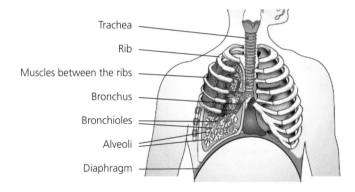

Trachea
Rib
Muscles between the ribs
Bronchus
Bronchioles
Alveoli
Diaphragm

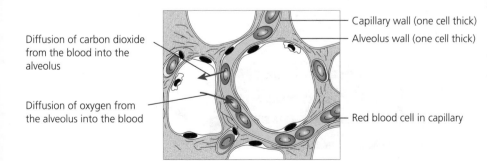

Diffusion of carbon dioxide from the blood into the alveolus

Diffusion of oxygen from the alveolus into the blood

Capillary wall (one cell thick)
Alveolus wall (one cell thick)

Red blood cell in capillary

Figure 11 Mammalian lungs

Exam tip

Don't say that alveoli have a thin 'cell wall'. A cell wall is a structure surrounding a plant cell.

Gas exchange in fish

Fish obtain their oxygen from the dissolved oxygen that is present in water. The amount of oxygen in a given volume of water is less than that in air and it decreases as temperature increases. Moreover, oxygen diffuses more slowly through water than through air. The gas exchange surface of a fish is the secondary lamellae of its gill filaments. The adaptations for efficient gas exchange include the following:

- All the secondary lamellae together have a huge surface area.
- The distance between the water flowing over the lamellae and the blood capillaries inside it is very small.
- Water is moved constantly over the gills, drawn in through the mouth and out beneath the operculum, providing a fresh supply of water containing relatively high concentrations of oxygen and low concentrations of carbon dioxide.
- Blood vessels carry oxygenated blood away from the gills, so maintaining a diffusion gradient.
- The flow of the water and the blood is in opposite directions — a counter-current flow — which helps to maintain a diffusion gradient across the whole surface of the lamellae (Figure 12).

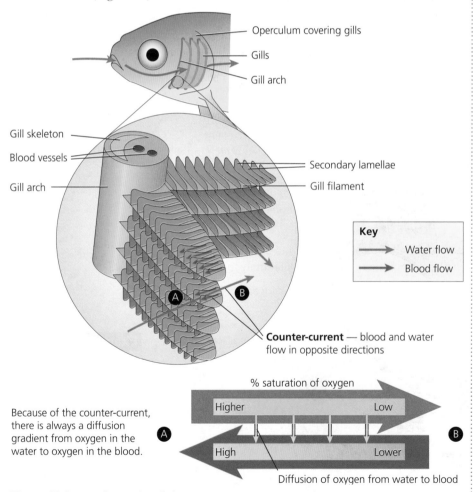

Figure 12 Gas exchange in a fish

Gas exchange in insects

Unlike mammals and fish, insects do not transport respiratory gases in their blood. Instead, they have a system of tubes called tracheae and tracheoles, which penetrate right inside body tissues. The gas exchange surface of an insect is the ends of the tracheoles. The adaptations for efficient gas exchange include the following:

- All the tracheoles together have a large surface area.
- The end walls of the tracheoles are very thin, with no thickening.
- The tracheoles penetrate right inside tissues so diffusion distances are kept short. In muscle tissue, they penetrate inside the cells and deliver oxygen directly to mitochondria.
- In some insects, breathing movements cause tidal air flow into and out of the tracheal system, helping to maintain diffusion gradients.

The position of the gas exchange surface deep inside the body helps to reduce water loss. The spiracles at the entrances to the tracheae can be closed completely if required.

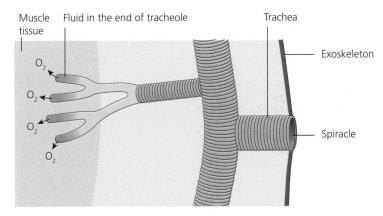

Figure 13 Gas exchange in an insect

Core practical 7

Dissect an insect to show the structure of the gas exchange system, taking into account the safe and ethical use of organisms

A locust, cockroach or other large insect can be dissected to show the tracheal system. Detailed instructions, including diagrams, can be found at
www.nuffieldfoundation.org/practical-biology/dissection-ventilation-system-locust

It is important that the insect is killed humanely and is not subjected to undue stress. One way of doing this is to place the insect into a container and then fill the container with carbon dioxide gas.

Wash your hands thoroughly after handling the insect. If you cut yourself with any of the dissection instruments, wash the wound thoroughly, put on some antiseptic cream and cover completely with a plaster. Dissection instruments should be sterilised before and after use.

Gas exchange in flowering plants

Plants have a large surface area to volume ratio because they have a branching shape. Gas exchange occurs:

■ across the walls of the mesophyll cells inside the leaf
■ through loosely packed groups of cells (lenticels) in the surface of the stem
■ through root surfaces

During daylight, leaf cells photosynthesise faster than they respire, and so take in carbon dioxide and give out oxygen. During darkness, leaf cells respire but do not photosynthesise, and so take in oxygen and give out carbon dioxide.

Note that plants do not use a transport system to move gases from the exchange surface. These gases move solely by diffusion (Figure 14).

Gas exchange in a leaf

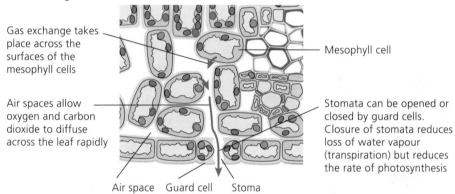

Gas exchange takes place across the surfaces of the mesophyll cells

Air spaces allow oxygen and carbon dioxide to diffuse across the leaf rapidly

Mesophyll cell

Stomata can be opened or closed by guard cells. Closure of stomata reduces loss of water vapour (transpiration) but reduces the rate of photosynthesis

Air space   Guard cell   Stoma

Structure of a lenticel

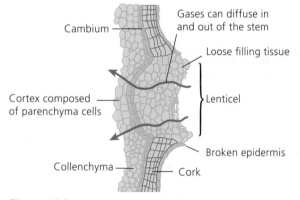

Cambium

Gases can diffuse in and out of the stem

Loose filling tissue

Lenticel

Cortex composed of parenchyma cells

Broken epidermis

Collenchyma   Cork

Figure 14 Gas exchange in a flowering plant

Summary

After studying this topic, you should be able to:
■ list the features of efficient gas exchange surfaces

■ explain how mammals, fish, insects and flowering plants are adapted for gas exchange
■ describe how to dissect an insect to show the structure of the gas exchange system

Circulation

Humans, like all mammals, have a circulatory system made up of a four-chambered heart and a system of blood vessels. The system is a double circulatory system in which the blood passes around the **pulmonary system** to the lungs and then back to the heart before passing around the **systemic system** to deliver oxygen to the rest of the body.

In contrast, fish have a single circulatory system in which the blood passes from the heart to the gills and then directly to the rest of the body without returning to the heart. This means that the blood arriving at the tissues is not at as high a pressure as in mammals and cannot deliver oxygen as efficiently.

The heart and circulatory system in mammals

The structure of the heart

The heart of a mammal has four chambers. The two atria receive blood and the two ventricles push blood out of the heart. The atria and ventricle on the left side of the heart contain oxygenated blood, whereas those on the right side contain deoxygenated blood (Figures 15 and 16). The walls of the heart are made of cardiac muscle.

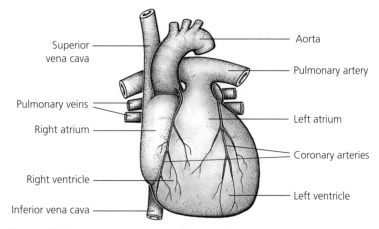

Figure 15 External view of a mammalian heart

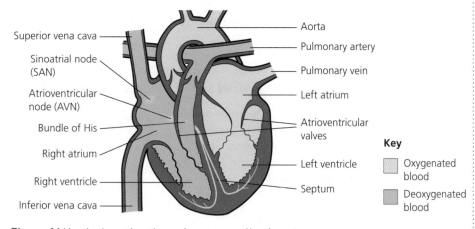

Figure 16 Vertical section through a mammalian heart

When muscle contracts, it gets shorter. Contraction of the cardiac muscle in the walls of the heart therefore causes the walls to squeeze inwards on the blood inside the heart. Both sides of the heart contract and relax together. The complete sequence of one heart beat is called the **cardiac cycle** (Figure 17).

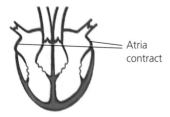

During **atrial systole**, the muscle in the walls of the atria contracts, pushing more blood into the ventricles.

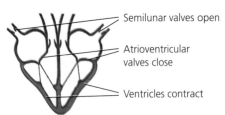

During **ventricular systole**, the muscle in the walls of the ventricles contracts. This causes the pressure of the blood inside the ventricles to become greater than in the atria, forcing the atrioventricular valves shut. The blood is forced out through the aorta and pulmonary artery.

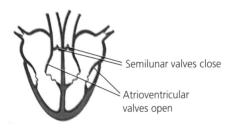

During **diastole**, the heart musles relax. The pressure inside the ventricles becomes less than inside the aorta and pulmonary artery, so the blood inside these vessels pushes the semilunar valves shut. Blood flows into the atria from the veins, so the cycle is ready to begin again.

Exam tip

Remember that the valves open and close because of differences in blood pressure on either side of them, as shown in Figure 18.

Figure 17 The cardiac cycle

Control of the cardiac cycle

Cardiac muscle is **myogenic** — that is, it contracts and relaxes automatically without the need of stimulation by nerves. The rhythmic, coordinated contraction of the cardiac muscle in different parts of the heart is coordinated through electrical impulses passing through the cardiac muscle tissue.

In the wall of the right atrium there is a patch of muscle tissue called the **sinoatrial node** (**SAN**). This has an intrinsic rate of contraction a little higher than that of the rest of the heart muscle. As the cells in the SAN contract, they generate action potentials (electrical impulses in nerve cells) that sweep along the muscle in the wall of the right and left atria. This causes the muscle to contract. This is **atrial systole**.

When the action potentials reach the **atrioventricular node** (**AVN**) in the septum, they are delayed briefly. They then sweep down the septum between the ventricles, along fibres in the **bundle of His** and then up through the ventricle walls. This causes the ventricles to contract slightly after the atria. The left and right ventricles contract together, from the bottom up. This is **ventricular systole**.

There is then a short delay before the next wave of action potentials is generated in the SAN. During this time, the heart muscles relax. This is **diastole**.

Exam tip

Remember that both sides of the heart contract and relax at the same time.

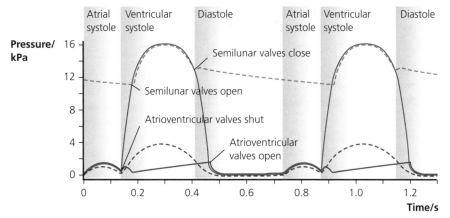

Notice that the valves open and close according to the relative pressure in the chambers that they separate. For example, when the pressure in the ventricle is greater than in the atrium, the net force of the blood is upward and this forces the atrioventricular valves shut.

Figure 18 Pressure changes during the cardiac cycle

This electrical activity in the heart can be picked up by electrodes placed on the chest. A recording of this activity is called an **electrocardiogram,** or **ECG** (Figure 19).

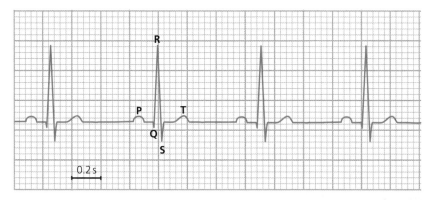

Figure 19 A normal ECG

Blood vessels

Arteries carry blood away from the heart. The blood that flows through them is pulsing and at a high pressure. They therefore have thick, elastic walls that can expand and recoil as the blood pulses through. The walls also contain variable amounts of smooth muscle. The arteries branch into smaller vessels called **arterioles**. These also contain smooth muscle in their walls, which can contract and make the lumen (space inside) smaller. This helps to control the flow of blood to different parts of the body. (Note that the muscle in the walls of arteries does not help to 'push' the blood through them.)

Veins carry low pressure blood back to the heart. Their walls do not need to be as tough or as elastic as those of arteries. The lumen is larger than in arteries, which reduces friction that would otherwise slow down blood movement. They contain

Knowledge check 12

Using Figure 18, calculate the number of heart beats in 1 minute.

Key
P   Electrical impulses spreading over atrium; atrial muscle contracts
QRS Electrical impulses spreading over ventricles; ventricular muscle contracts
T   Ventricular muscle recovering; muscle relaxes

valves to ensure that the blood does not flow the wrong way. Blood is kept moving through many veins, such as those in the legs, by the squeezing effect produced by the contraction of the body muscles close to them, which are used when walking.

Capillaries are tiny vessels with just enough space for red blood cells to squeeze through. Their walls are only one cell thick and there are often gaps in the walls through which plasma (the liquid component of blood) can leak out. Capillaries deliver nutrients, hormones and other requirements to body cells and take away their waste products. Their small size and thin walls minimise diffusion distance, enabling exchange to take place rapidly between the blood and the body cells.

Figure 20 shows these three types of blood vessel and Figure 21 shows the main blood vessels in the human body.

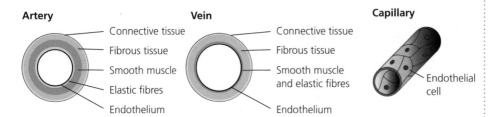

Figure 20 The structure of blood vessels

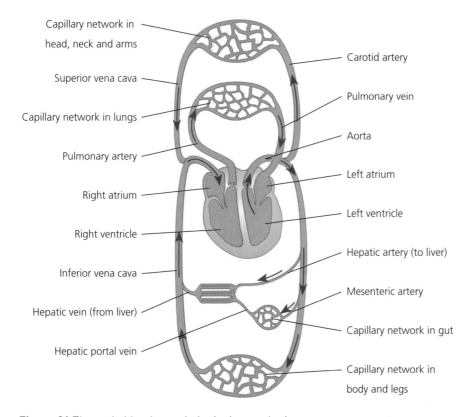

Figure 21 The main blood vessels in the human body

Knowledge check 13

Do all arteries carry oxygenated blood? Explain your answer.

Pressure changes in the circulatory system

The pressure of the blood changes as it moves through the circulatory system.

In the arteries, blood is at high pressure because it has just been pumped out of the heart. The pressure oscillates (goes up and down) in time with the heart beat. The stretching and recoil of the artery walls helps to smooth the oscillations, so the pressure becomes gradually steadier the further the blood moves along the arteries. The mean pressure also gradually decreases.

In the veins, blood is at a low pressure as it is now a long way from the pumping effect of the heart.

The total cross-sectional area of the capillaries is greater than that of the arteries that supply them, so blood pressure is less inside the capillaries than inside arteries.

Figure 22 shows the changes in the pressure of the blood as it flows through the circulatory system.

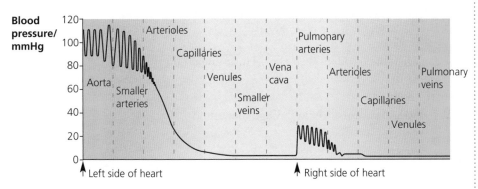

Figure 22 Pressure changes in the circulatory system

Knowledge check 14

Why is the pressure produced by the left side of the heart greater than that produced by the right side of the heart?

Blood

Blood contains cells floating in a liquid called **plasma**.

Erythrocytes (red blood cells) transport oxygen from lungs to respiring tissues. They are very small and have the shape of a biconcave disc. This increases their surface area to volume ratio, allowing rapid diffusion of oxygen into and out of them. They contain haemoglobin, which combines with oxygen to form oxyhaemoglobin in areas of high concentration (such as the lungs) and releases oxygen in areas of low concentration (such as respiring tissues). They do not contain a nucleus or mitochondria.

Leucocytes (white blood cells) include monocytes, neutrophils, lymphocytes and eosinophils. **Monocytes** and **neutrophils** are phagocytic cells. They engulf and digest unwanted cells, such as damaged body cells or pathogens. They are larger than red blood cells. **Lymphocytes** respond to particular pathogens by secreting antibodies or by directly destroying them. Each lymphocyte is able to recognise one particular pathogen and respond to it by secreting one particular type of antibody or by attacking it. **Eosinophils** look like neutrophils but have fewer, redder granules in their cytoplasm. They respond to invasion by parasites by producing chemicals called cytokines, which activate other leucocytes.

Blood also contains small cell fragments called **platelets**, which are involved in blood clotting. Some examples of blood cells are shown in Figure 23.

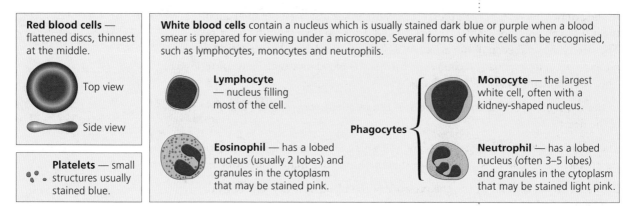

Figure 23 Some types of blood cells

The components of tissue fluid and lymph are described on pages 39–40.

Blood clotting

When a blood vessel is broken or damaged, the blood inside it clots. This is important so that:

- not too much blood is lost from the body
- pathogens (disease-causing microorganisms) cannot get in through the wound

The process is as follows:

1 Platelets stick together to form a plug in the wound.
2 Platelets and the damaged tissue in the blood vessel wall release a mixture of substances called **thromboplastin**.
3 Platelets also release calcium ions and other substances called **clotting factors**.
4 Blood plasma contains a soluble protein called **prothrombin**. Thromboplastin causes prothrombin to change to an enzyme called **thrombin**.
5 Blood plasma also contains a soluble protein called **fibrinogen**. In the presence of calcium ions, thrombin causes fibrinogen to change to an insoluble fibrous protein called **fibrin**.
6 Fibrin precipitates to form long fibres. Platelets and red blood cells get tangled in the fibres and form a blood clot.

Atherosclerosis

Atherosclerosis is also known as 'hardening of the arteries'. It develops as the result of this sequence of events:

- The endothelium (inner layer of tissue) of an artery gets damaged and becomes rough. This happens to all of us as we age, but it tends to happen earlier in people with high blood pressure. It is also caused by chemicals in cigarette smoke.
- The damaged tissues and the white cells in the blood respond to the damage by secreting several different chemicals. White blood cells crawl out of the blood and into the artery wall. Cholesterol from the blood also builds up in the wall, forming a deposit called an **atheroma**.

Knowledge check 15

Which of the substances involved in blood clotting are proteins?

- Over time, fibrous tissue builds up around the cholesterol, forming a **plaque** (Figure 24).

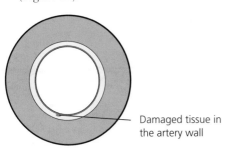
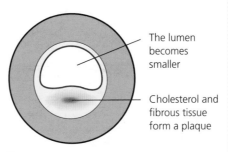

Damaged tissue in the artery wall

The lumen becomes smaller

Cholesterol and fibrous tissue form a plaque

Figure 24 Formation of plaque in an artery wall

- The plaque makes the artery wall thicker and less elastic. The artery is therefore narrower so it is more difficult for blood to flow through it. Blood pressure increases. The high blood pressure increases the risk that more plaques will form.
- The slow flow of blood past a plaque means platelets come into close contact with the damaged tissue in the vessel wall. This can stimulate a blood clot to form. When this happens inside a blood vessel, it is called a **thrombosis**. The clot can block the artery.

If a blood clot forms in a coronary artery, the flow of blood to part of the heart muscle is stopped. The muscle no longer gets a supply of oxygen and nutrients, so cannot respire aerobically. It therefore has no energy to contract and may even die. This part of the heart stops working and the person has a heart attack. The severity of the attack depends partly on how much muscle is affected. Atherosclerosis also increases the risk of stroke — damage to the brain caused by a burst or blocked blood vessel.

The following factors increase the risk that people will develop atherosclerosis:

- having genes that predispose them to it
- being overweight
- being male
- eating a high-fat diet, particularly one rich in saturated fats
- smoking cigarettes
- doing little physical exercise

The risk of developing cardiovascular disease also increases with age.

Transport of gases in the blood

Haemoglobin and oxygen transport

Haemoglobin (Hb) is a protein with a quaternary structure. A haemoglobin molecule is made up of four polypeptide chains, each of which has a haem group at its centre. Each haem group contains an Fe^{2+} ion that is able to combine reversibly with oxygen, forming **oxyhaemoglobin**. Each iron ion can combine with two oxygen atoms, so one haemoglobin molecule can combine with eight oxygen atoms.

Oxygen concentration can be measured as partial pressure, in kilopascals (kPa). Haemoglobin combines with more oxygen at high partial pressures than it does at low partial pressures. At high partial pressures of oxygen, all the haemoglobin will

Exam tip

Notice that the plaque is inside the wall of the artery, not just stuck on its inner surface.

Knowledge check 16

It may soon become possible to have your DNA analysed to see if you have genes increasing the risk of cardiovascular disease. Suggest the benefits and potential problems of this.

Exam tip

It is surprisingly common for examination candidates to confuse red blood cells with haemoglobin molecules — make sure you are not one of those candidates.

be combined with oxygen and we say that it is 100% saturated with oxygen. A graph showing the relationship between the partial pressure of oxygen and the percentage saturation of haemoglobin with oxygen is known as an **oxygen dissociation curve** (Figure 25).

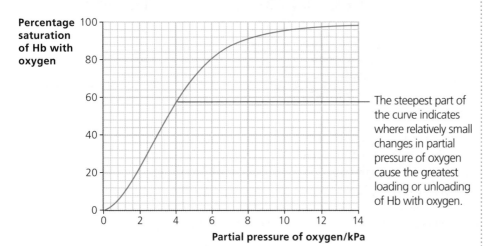

The steepest part of the curve indicates where relatively small changes in partial pressure of oxygen cause the greatest loading or unloading of Hb with oxygen.

Figure 25 The oxygen dissociation curve for haemoglobin

In the lungs, the partial pressure of oxygen may be around 12 kPa. You can see from Figure 25 that the haemoglobin will be about 98% saturated. In a respiring muscle, the partial pressure of oxygen may be around 2 kPa. The haemoglobin will be about 23% saturated. Therefore, when haemoglobin from the lungs arrives at a respiring muscle it gives up more than 70% of the oxygen it is carrying.

The Bohr effect

The presence of carbon dioxide increases acidity, that is the concentration of H^+ ions. When this happens, the haemoglobin combines with H^+ ions and releases oxygen.

Red blood cells contain an enzyme called carbonic anhydrase, which catalyses the reaction of carbon dioxide and water to form carbonic acid:

$$CO_2 + H_2O \rightleftharpoons H_2CO_3$$

The carbonic acid then dissociates:

$$H_2CO_3 \rightleftharpoons H^+ + HCO_3^-$$

The hydrogen ions combine with haemoglobin to form haemoglobinic acid. This causes the haemoglobin to release oxygen.

Therefore, in areas of high carbon dioxide concentration, haemoglobin is less saturated with oxygen than it would be if there were no carbon dioxide present. This is called the **Bohr effect** (Figure 26). It is useful in enabling haemoglobin to unload more of its oxygen in tissues where respiration (which produces carbon dioxide) is taking place.

Myoglobin

Myoglobin has a similar structure to one of the four polypeptide chains that make a haemoglobin molecule. Myoglobin has a high affinity for oxygen. It binds with it

Exam tip

When explaining how the structure of haemoglobin is related to its functions, remember that its ability to unload oxygen is just as important as its ability to combine with it.

easily, but does not dissociate unless the partial pressure of oxygen is very low. This makes it an ideal oxygen store in muscle cells.

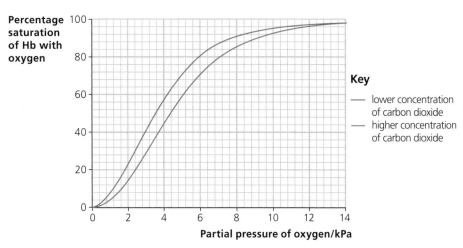

Figure 26 The Bohr effect

Fetal haemoglobin

A fetus obtains its oxygen by diffusion from its mother's blood, through the placenta. The haemoglobin of a fetus has a dissociation curve that lies to the left of the curve for adult haemoglobin. This means that, at any particular partial pressure of oxygen, fetal haemoglobin has a higher affinity for oxygen than adult haemoglobin. Oxygen therefore moves readily from the mother's haemoglobin to that of her fetus.

Transfer of materials between the circulatory system and cells

Tissue fluid and lymph

Capillaries have tiny gaps between the cells in their walls. Near the arteriole end of a capillary, there is relatively high hydrostatic pressure inside the capillary and some components of the plasma leak out through these gaps to fill the spaces between the body cells. This leaked plasma is called **tissue fluid**.

Tissue fluid is therefore similar to blood plasma. However, large molecules such as albumin (a protein carried in solution in blood plasma) and other plasma proteins cannot get through the pores and so remain in the blood plasma.

The tissue fluid bathes the body's cells. Substances such as oxygen, glucose or urea can move between the blood plasma and the cells by diffusing through the tissue fluid. Some tissue fluid moves back into the capillaries, becoming part of the blood plasma once more. This happens especially at the venule end of the capillary, where the hydrostatic pressure of the blood is lower, producing a pressure gradient down which the tissue fluid can flow. Moreover, the water potential of the tissue fluid is higher than that of the blood at this end of the capillary because of the earlier loss of water from the capillary and the high concentration of plasma proteins. Water therefore moves down this water potential gradient from the tissue fluid into the blood.

However, not all of the fluid returns to the blood in this way. Some of the tissue fluid collects into blind-ending vessels called lymphatic vessels. It is then called **lymph**. Lymphatic vessels have valves that allow fluid to flow into them and along them, but not back out again. They carry the lymph towards the subclavian veins (near the collarbone), where it is returned to the blood. The lymph passes through lymphatic glands where white blood cells accumulate. Lymph therefore tends to carry higher densities of white blood cells than are found in blood plasma or tissue fluid.

Summary

After studying this topic, you should be able to:
- describe the structure of the heart, arteries, veins and capillaries
- explain the advantages of a double circulatory system
- describe and explain the cardiac cycle, including the roles of the SAN, AVN and bundles of His
- interpret ECG traces
- describe the different types of cells found in blood and outline their functions
- describe how blood clotting occurs, including the role of platelets
- describe how atherosclerosis happens, its effects on health and risk factors
- explain how haemoglobin transports oxygen, including the Bohr effect
- explain oxygen dissociation curves for haemoglobin, myoglobin and fetal haemoglobin
- explain the formation and roles of tissue fluid and lymph

Transport in plants

Plants have two transport systems:
- **xylem**, which transports water and inorganic ions from the roots to all other parts of the plant
- **phloem**, which transports substances made in the plant, such as sucrose and amino acids, to all parts of the plant

Xylem tissue contains dead, empty cells with no end walls. These are called **xylem vessel elements**. They are arranged in long lines to form **xylem vessels** (Figure 27). These are long, hollow tubes through which water moves by mass flow from the roots to all other parts of the plant.

Transverse section

Cell wall containing cellulose and lignin — lignin makes the wall impermeable to water and provides strength, so the vessel element does not collapse when there is negative pressure inside it

Narrow lumen increases area of water in contact with wall; water molecules adhere to the walls and this helps to prevent breakage of the water column

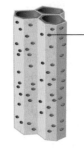

Pit in cell wall allows movement of water out of the vessel element to other vessel elements or to neighbouring tissues

Longitudinal section

Dead cells have no contents, allowing easy movement of a continuous column of water by mass flow

Loss of end walls allows continuous movement of a column of water by mass flow

Figure 27 The structure of xylem vessels

Transport in xylem

Figure 28 shows the pathway taken by water through a plant.

The driving force that causes this movement is the loss of water vapour from the leaves. This is called **transpiration**.

Knowledge check 17

Use the diagram at the bottom left-hand side of Figure 28 to draw a plan diagram of a transverse section of a root.

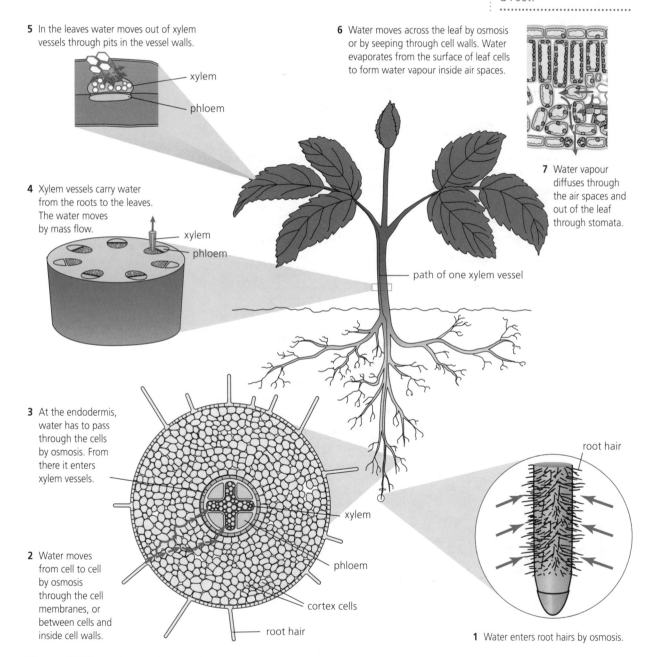

5 In the leaves water moves out of xylem vessels through pits in the vessel walls.

xylem

phloem

6 Water moves across the leaf by osmosis or by seeping through cell walls. Water evaporates from the surface of leaf cells to form water vapour inside air spaces.

7 Water vapour diffuses through the air spaces and out of the leaf through stomata.

4 Xylem vessels carry water from the roots to the leaves. The water moves by mass flow.

xylem

phloem

path of one xylem vessel

3 At the endodermis, water has to pass through the cells by osmosis. From there it enters xylem vessels.

root hair

xylem

phloem

cortex cells

root hair

2 Water moves from cell to cell by osmosis through the cell membranes, or between cells and inside cell walls.

1 Water enters root hairs by osmosis.

Figure 28 The pathway of water through a plant

How water moves from soil to air

Water moves from the soil to the air through a plant down a water potential gradient. The water potential in the soil is generally higher than in the air. The water potential in the leaves is kept lower than the water potential in the soil because of the loss of water vapour by transpiration. Transpiration maintains the water potential gradient.

Water enters root hair cells by osmosis, moving down a water potential gradient from the water in the spaces between soil particles, through the cell-surface membrane and into the cytoplasm and vacuole of the root hair cell. The water then moves from the root hair cell to a neighbouring cell by osmosis, down a water potential gradient. This is called the **symplastic pathway**.

Water also seeps into the cell wall of the root hair cell. This does not involve osmosis, as no partially permeable membrane is crossed. The water then seeps into and along the cell walls of neighbouring cells. This is called the **apoplastic pathway**. In most plant roots, the apoplastic pathway carries more water than the symplastic pathway.

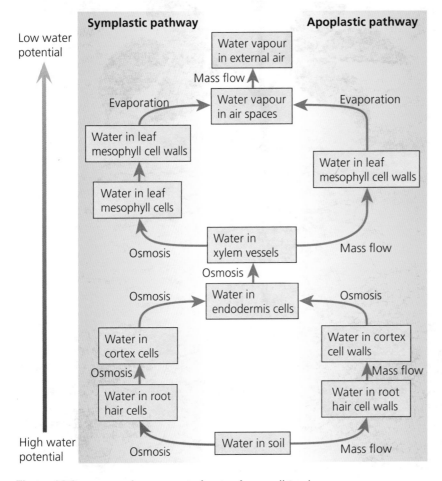

Figure 29 Summary of movement of water from soil to air

When the water nears the centre of the root, it encounters a cylinder of cells called the **endodermis**. Each cell has a ring of impermeable **suberin** around it, forming the **Casparian strip**. This prevents water continuing to seep through cell walls. It therefore travels through these cells by the symplastic pathway.

The water moves into the xylem vessels from the endodermis. Water moves up the xylem vessels by **mass flow** — that is, in a similar way to water flowing in a river. The water molecules are held together by hydrogen bonds between them, keeping the water column unbroken. This is called **cohesion** and it helps to create a **tension** in the water column as it is drawn upwards. This is known as the cohesion–tension model. The water molecules are also attracted to the cellulose and lignin in the walls of the xylem vessels by a force called **adhesion**. There is a relatively low hydrostatic pressure at the top of the column, produced by the loss of water by transpiration. This lowering of hydrostatic pressure causes a pressure gradient from the base to the top of the xylem vessel.

In a leaf, water moves out of xylem vessels through pits and then across the leaf by the symplastic and apoplastic pathways. Water evaporates from the wet cell walls into the leaf spaces and then diffuses out through the stomata.

Transpiration

Transpiration is the loss of water vapour from a plant. Most transpiration happens in the leaves. A leaf contains many cells in contact with air spaces in the mesophyll layers. Liquid water in the cell walls changes to water vapour, which diffuses into the air spaces. The water vapour then diffuses out of the leaf through the stomata (down a water potential gradient) and into the air surrounding the leaf.

> **Exam tip**
>
> Remember that it is water vapour that diffuses out of a plant leaf, not liquid water. The change from liquid to vapour takes place at the surface of the cell walls inside the mesophyll layers of the leaf.

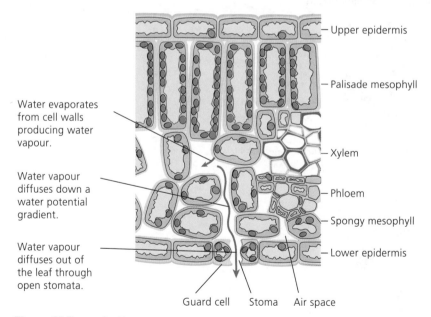

Water evaporates from cell walls producing water vapour.

Water vapour diffuses down a water potential gradient.

Water vapour diffuses out of the leaf through open stomata.

— Upper epidermis

— Palisade mesophyll

— Xylem

— Phloem

— Spongy mesophyll

— Lower epidermis

Guard cell   Stoma   Air space

Figure 30 Transpiration

Each stoma is surrounded by a pair of guard cells. These can change shape to open or close the stoma. In order to photosynthesise, the stomata must be open so that carbon dioxide can diffuse into the leaf. Plants cannot therefore avoid losing water vapour by transpiration.

Transpiration is affected by several factors:

- **High temperature** increases the rate of transpiration. This is because at higher temperatures water molecules have more kinetic energy. Evaporation from the cell walls inside the leaf therefore happens more rapidly and diffusion also happens more rapidly.
- **High humidity** decreases the rate of transpiration. This is because the water potential gradient between the air spaces inside the leaf and the air outside is less steep, so diffusion of water vapour out of the leaf happens more slowly.
- **High wind speed** increases the rate of transpiration. This is because the moving air carries away water vapour from the surface of the leaf, helping to maintain a water potential gradient between the air spaces inside the leaf and the air outside.
- **High light intensity** may increase the rate of transpiration. This is because the plant may be photosynthesising rapidly, requiring a rapid supply of carbon dioxide. This means that more stomata are likely to be open, through which water vapour can diffuse out of the leaf.

Transport in phloem

Phloem tissue contains cells called **sieve tube elements**. Unlike xylem vessel elements, these are living cells and contain cytoplasm and a few organelles but no nucleus. Their walls are made of cellulose. A **companion cell** is associated with each sieve tube element.

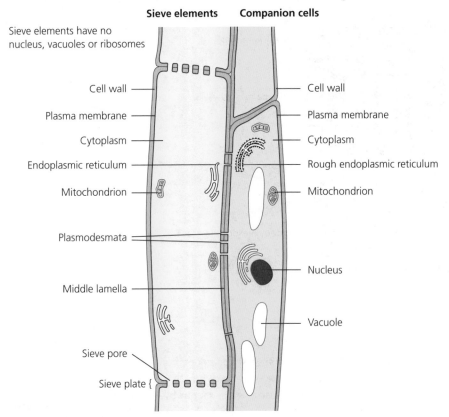

Figure 31 Phloem tissue

> **Exam tip**
>
> Candidates often write 'Temperature increases the rate of transpiration'. This would not get a mark because you need to say that increased temperature increases the rate of transpiration.

How assimilates move through phloem tissue

Substances made by a plant, called **assimilates**, are transported in phloem tissue. This process is called **translocation**. A part of a plant where assimilates such as sucrose enter the phloem is called a **source**. A part where assimilates leave the phloem is called a **sink**. For example, a leaf may be a source and a root may be a sink.

Translocation of sucrose and other assimilates is an energy-requiring process. Respiration in companion cells at a source provides ATP that is used to fuel the transport of sucrose into the companion cell. This is done by pumping protons (hydrogen ions) out of the companion cell. As the protons diffuse back into the companion cell, they pass through a co-transporter protein that allows both protons and sucrose molecules to enter the cell. The protons carry sucrose molecules with them. This increases the concentration of sucrose in the companion cell, so that it moves by diffusion down a concentration gradient, through the plasmodesmata and into the phloem sieve element. The increased concentration of sucrose in the companion cell and phloem sieve element produces a water potential gradient from the surrounding cells into the companion cell and phloem sieve element. Water moves down this gradient.

At a sink, sucrose diffuses out of the phloem sieve element and down a concentration gradient into a cell that is using sucrose. This produces a water potential gradient, so water also diffuses out of the phloem sieve element.

The addition of water at the source and the loss of water at the sink produces a higher hydrostatic pressure inside the phloem sieve element at the source than at the sink. Phloem sap therefore moves by mass flow down this pressure gradient, through the phloem sieve elements and through the sieve pores, from source to sink. There is still uncertainty about the precise details of this mechanism of transport and so it is known as the mass flow hypothesis. However, virtually all the evidence so far collected supports this theory, for example:

- When a stem is cut, the contents of the phloem leak out rapidly, indicating that they are under pressure.
- Measurements of concentrations of solutes at sources and sinks show that there is a concentration gradient for them from source to sink.

Evidence against this theory includes:

- The presence of companion cells indicates that transport in phloem is an active process rather than relying on mass flow. However, this is accounted for by the need for active loading of sucrose into the phloem tissue.
- Assimilates have been shown to move both ways in phloem tissue at the same time. However, it has so far not been shown that this bidirectional movement occurs in the same sieve tube; it seems to happen only in different sieve tubes and so does not disprove the mass flow hypothesis.

Knowledge check 18

At which of these stages of transport of sucrose does the plant have to provide energy?
- Loading sucrose into the phloem sieve tube.
- Mass flow of phloem sap from source to sink.

Knowledge check 19

Explain why the contents of xylem vessels always flow upwards from roots to leaves, whereas the contents of phloem sieve tubes can flow either upwards or downwards.

Knowledge check 20

Plants contain many different organs, including flowers, leaves, roots and fruits. In a temperate climate, different organs in a plant are sources and sinks at different times of the year. State which of these organs will be sources and which will be sinks:

a in spring, when the leaves are just starting to grow

b in summer, when there is plenty of sunlight and flower buds are beginning to open

c in autumn, when there is still plenty of sunlight and the plant is building up stores of starch in its roots ready for the winter; flowers have been fertilised and fruits are developing

d in winter, when there is little sunlight, and the plant relies on starch stores in its roots

Core practical 8

Investigate factors affecting water uptake by plant shoots using a potometer

It is difficult to measure the rate at which water vapour is lost from leaves. It is much easier to measure the rate at which a plant, or part of a plant, takes up water. Most of the water taken up is lost through transpiration, so we can generally assume that an increase in the rate of take-up of water indicates an increase in the rate of transpiration.

The apparatus used to measure the rate of take-up of water of a plant shoot is called a potometer. This can simply be a long glass tube. More complex potometers may have reservoirs, which make it easier to refill the tube with water, or a scale marked on them.

Fix a short length of rubber tubing over one end of the long glass tube. Completely submerge the tube in water. Move it around to get rid of all the air inside it and fill it with water. Make absolutely sure there are no air bubbles.

Take a leafy shoot from a plant and submerge it in the water alongside the glass tube. Using a sharp blade, make a slanting cut across the stem.

Push the cut end of the stem into the rubber tubing. Make sure the fit is tight and that there are no air bubbles. If necessary, use a small piece of wire to fasten the tube tightly around the stem.

Take the whole apparatus out of the water and support it upright. Wait at least ten minutes for it to dry out. If the glass tube is not marked with a scale, place a ruler or graph paper behind it.

continued

Start a stopwatch and read the position of the air/water meniscus (which will be near the base of the tube). Record its position every 2 minutes (or whatever time interval seems sensible). Stop when you have ten readings or when the meniscus is one-third of the way up the tube.

Change the environmental conditions and continue to take readings. For example, you could use a fan to increase 'wind speed' or move the apparatus into an area where the temperature is higher or lower.

Plot distance moved by the meniscus against time for each set of readings on the same axes. Draw best-fit lines. Calculate the mean distance moved per minute or calculate the slope of each line. This can be considered to be the rate of transpiration.

Summary

After studying this topic, you should be able to:

- describe the structure of xylem tissue and phloem tissue, and relate their structures to their functions
- describe the symplastic and apoplastic pathways for the movement of water across tissues
- explain the cohesion–tension model
- explain how temperature, humidity, air movement and light affect rates of transpiration
- describe the mass flow hypothesis for translocation in phloem and outline its strengths and weaknesses
- describe how to investigate factors that affect water uptake by plant shoots using a potometer

Questions & Answers

In this section there are two sample examination papers that contain questions in styles similar to those in the Edexcel AS Paper 2 and A-level Papers 1 and 2 of the Biology B specification. However, whereas the A-level papers test content from all or most of the topics you will study during your course, these sample papers test only content from Topics 3 and 4.

You have 1 hour 30 minutes to do each sample paper. There are 80 marks on the paper, so you can spend 1 minute per mark, plus time for reading the questions and checking your answers. If you find you are spending too long on one question, move on to another that you can answer more quickly. If you have time at the end, come back to the difficult one.

Some of the questions require you to recall information that you have learned. Be guided by the number of marks awarded to suggest how much detail you should give in your answer. The more marks there are, the more information you need to give.

Some of the questions require you to use your knowledge and understanding in new situations. Don't be surprised to find something completely new in a question — something you have not seen before. Just think carefully about it and find something that you do know that will help you to answer it.

Do think carefully before you begin to write — the best answers are short and relevant. If you target your answer well, you can get many marks for a small amount of writing. Don't ramble on and say the same thing several times over or wander off into answers that have nothing to do with the question. As a general rule, there will be twice as many answer lines as marks. So you should try to answer a 3-mark question in no more than six lines of writing. If you are writing much more than that, you almost certainly have not focused your answer tightly enough.

Look carefully at exactly what each question wants you to do. For example, if it asks you to 'Explain', then you need to say *how* or *why* something happens, not just what happens. Many students lose large numbers of marks by not reading the question carefully.

Each question is followed by a brief analysis of what to watch out for when answering the question (icon ⒠). All student responses are then followed by comments. These are preceded by the icon ⒠ and indicate where credit is due. In the weaker answers, they also point out areas for improvement, specific problems and common errors, such as lack of clarity, weak or non-existent development, irrelevance, misinterpretation of the question and mistaken meanings of terms.

■ Sample paper 1

Question 1

(a) A species can be defined as a group of organisms with similar characteristics that interbreed to produce fertile offspring. State two reasons why it may be difficult to assign organisms to different species using this definition. (2 marks)

A group of islands contains three species of mice, with each species found on only one island. A fourth species is found on the mainland. A region of the DNA of each species was sequenced and the percentage differences between the samples were calculated. The results are shown in Table 1.

	Mainland	Island A	Island B
Island A	6.1	–	–
Island B	4.8	9.7	–
Island C	5.2	10.3	7.5

Table 1

(b) Discuss how these results suggest that the species of mouse on each island has evolved from the species on the mainland and not from one of the other island species. (5 marks)

(c) Each of these four species of mice is unable to breed with any of the other species, even if they are placed together. Suggest how reproductive isolation between the mice could have arisen and explain its role in speciation. (6 marks)

Total: 13 marks

ⓔ Note that part (c) asks you to do two things — first 'suggest' and then 'explain'.

Student A

(a) The organisms might not live close to one another, so you might not be able to know whether they can interbreed or not ✓. They might be fossils ✓ so you cannot ever know anything about how they reproduced.

ⓔ **2/2 marks awarded** These are two suitable suggestions.

(b) The three island mice each have DNA more similar to the mainland mouse than to each other ✓✓, so they have probably all evolved from the mainland mouse. If one of the island mice had evolved from another island mouse, their DNA would be more similar ✓.

ⓔ **3/5 marks awarded** This answer shows that student A has managed to work out what the table shows, but it does not provide enough detail in the discussion to get all the marks available.

(c) If they are on different islands, they will have different selection pressures ✓ so the mice might end up different. They might have different courtship behaviour ✓, so they won't be able to mate with each other ✓. You have to get reproductive isolation to produce a new species.

ⓔ **3/6 marks awarded** This is a reasonable description. The last sentence is moving towards another mark, but it really only repeats what is said in the question.

Student B

(a) Some organisms, such as many plants, only reproduce asexually, so they never breed with any other individual ✓. If two similar-looking organisms live in two different parts of the world, they cannot interbreed just because they are nowhere near each other, so we cannot know if they would interbreed if they lived in the same place ✓.

ⓔ **2/2 marks awarded** Both suggestions are suitable.

(b) From the table, we can see that each island mouse's DNA is more similar to the DNA of the mainland mouse than to any of the other island mice ✓. For example, the island C mice have DNA that is 10.3% different from the island A mice and 7.5% different from the island B mice, but only 5.2% different from the mainland mice ✓. The longer ago two species split away from each other, the more different we would expect their DNA to be ✓. This is because the longer the time, the more mutations ✓ are likely to have occurred, so there will be different base sequences ✓ in the DNA.

ⓔ **5/5 marks awarded** This is a good answer to a difficult question. Student B has made good use of the data in the table and has used some of the figures to support their answer. The last part of the answer explains why differences in DNA base sequences indicate the degree of relationship.

(c) A species is defined as a group of organisms that can interbreed with each other to produce fertile offspring ✓. So to get new species, you need to have something that stops them reproducing together so genes cannot flow ✓ from one species to the other. This might happen if different selection pressures ✓ acted on two populations of a species, so that different alleles were selected for ✓ and over many generations their genomes became more different ✓. This means they might be the wrong size and shape ✓ to be able to breed with each other.

ⓔ **6/6 marks awarded** The answer begins with a clearly explained link between speciation and reproductive isolation, and then goes on to describe how two populations could become reproductively isolated.

Question 2

Figure 1 shows the structure of the gas exchange system of an insect.

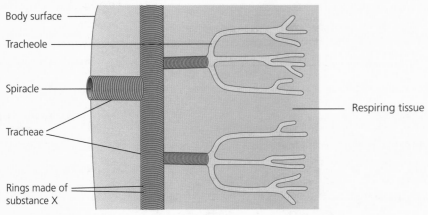

Figure 1

(a) What is substance X? (1 mark)

    A Cartilage   B Cellulose   C Chitin   D Lignin

(b) Explain how gases are exchanged between the environment and the respiring tissues. (3 marks)

Figure 2 shows how oxygen concentration in the atmosphere is thought to have varied over the last 550 million years.

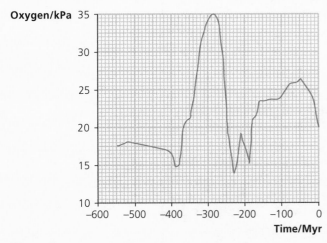

Figure 2

(c) (i) Calculate the percentage difference between the maximum partial pressure of oxygen over this time period and the partial pressure at the present time. Show your working. (3 marks)

(ii) Between about 400 and 300 million years ago, giant insects existed. Some of these were up to ten times larger than insects on earth today.

With reference to the data in the graph and your knowledge of the effects of surface area to volume ratio on gas exchange, suggest why larger insects were able to evolve during this time period.

(6 marks)

Total: 13 marks

ⓔ Take special care over part (c) (ii), where there are 6 marks available. Plan your answer before you begin to write.

Student A

(a) A ✗

ⓔ **0/1 mark awarded** Student A may have confused the rings of chitin in the insect tracheoles with rings of cartilage in the mammalian trachea.

(b) The insect opens and closes its spiracles to push air in and out of the trachea ✗. Oxygen diffuses from the end of the tracheoles into the respiring tissues down a concentration gradient ✓, and carbon dioxide diffuses the other way. The ends of the tracheoles have got liquid in them, which speeds up the diffusion ✗.

ⓔ **1/3 marks awarded** Although it is true that the spiracles can be opened or closed, this does not help to move air into or out of the tracheal system. The student correctly mentions a diffusion gradient, but does not explain what causes this. The statement about the liquid speeding up diffusion is incorrect; diffusion happens more slowly in liquids than in gases.

(c) (i) Maximum concentration was 35 kPa. Today it is 20 kPa. So the percentage difference is:

$$\frac{35}{20} \times 100$$
$$= 175\%$$

ⓔ **0/3 marks awarded** The student has correctly read the relevant figures from the graph, but has not found the difference between them and therefore has not calculated the percentage difference.

(ii) The graph shows that there was more oxygen in the air between 400 and 300 million years ago than there is today, or before that ✓. A big animal has a smaller surface area to volume ratio than a smaller one ✓, so giant insects would have had smaller surface areas compared to small insects ✗. This would make it difficult for them to get enough oxygen through their tracheal systems, but because there was more oxygen then they were able to get bigger ✓ and still be able to respire.

ⓔ 3/6 marks awarded The student has correctly identified a link between the high concentrations of oxygen in the atmosphere and the ability of large insects to survive. There is a lack of precision of language in the second sentence — in the statement that giant insects would have had smaller surface areas, this should refer to smaller surface area to volume ratios. There is no reference to how these large insects might have 'evolved', a term that is used in the question.

Student B

(a) C ✓

ⓔ 1/1 mark awarded The correct answer gains 1 mark.

(b) In some insects, the contraction and relaxation of muscles in the thorax during flight cause pressure changes in the tracheae, which act like breathing movements and move air into and out of the tracheal system ✓. Near the ends of the tracheoles, however, all movement is by diffusion ✓. There is a low concentration of oxygen in the tissues because they are respiring and using up oxygen ✓, so oxygen diffuses across the thin walls of the tracheole endings, down a diffusion gradient ✓. Carbon dioxide diffuses the other way.

ⓔ 3/3 marks awarded A clearly organised and fully correct answer.

(c) (i) 15 – 20 = 15 ✓. So percentage difference =
$$\frac{15}{35} \times 100 \checkmark$$
$$= 42.9\% \checkmark$$

ⓔ 3/3 marks awarded This is all correct and shown clearly.

(ii) There was much more oxygen in the atmosphere in the period 400 to 300 million years ago than there was before or since ✓. Insects are limited in the maximum size to which they can grow because they have external skeletons that are unable to support large bodies ✓. Also, they have an internal gas exchange system that delivers gases directly to the tissues. They need a relatively large surface area to volume ratio for this to work effectively. As animals get larger, their surface area to volume ratio decreases ✓, so large insects might not have sufficient gas exchange surface for them to obtain enough oxygen through their spiracles ✓. Also, the length of the tracheae would increase, so that gases have to travel further to get to the exchange surface ✓. However, when there is more oxygen in the air they could probably cope with being larger. Larger insects might have evolved then because they would be more able to resist predators ✓ or to catch prey. They would have had a selective advantage over smaller insects.

ⓔ 6/6 marks awarded This is a good answer, but perhaps a little more wordy than necessary. The student has identified a number of relevant points and expressed them carefully using the correct technical language. He/she has also referred to evolution — more could have been made of this strand of the question, with reference to natural selection, but as it is there is enough here for full marks.

Question 3

A student carried out an investigation into the effect of temperature on the permeability of the cell-surface membrane of beetroot cells. She decided to measure permeability by using a colorimeter to measure the absorbance of green light by the solutions in which samples of beetroot had been immersed.

(a) Describe how she could carry out this investigation. (5 marks)

(b) Suggest why she measured the absorbance of green light, not red light. (1 mark)

Figure 3 shows her results.

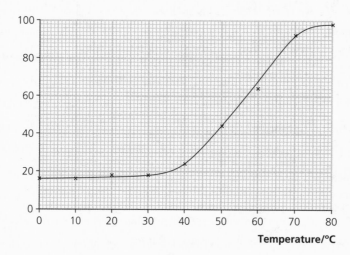

Figure 3

(c) (i) With reference to Figure 3, describe the effect of temperature on the absorbance of light in the colorimeter. (3 marks)

(ii) With reference to the structure of cell membranes, explain the effects you have described in (i). (4 marks)

Total: 13 marks

ⓔ Part (a) requires a description of experimental procedure. Make sure your answer is quite detailed — mention numbers, temperatures and times. When describing what is shown by a graph, outline the general trend and then home in on places where the gradient changes. Quote particular coordinates.

Student A

(a) She should cut up some pieces of beetroot, all the same size ✓. Put them into water at different temperatures ✓ and leave them for the same time ✓. Take the beetroot pieces out. Put the water in a colorimeter tube and measure how much light goes through it ✓. The less light goes through, the more the membrane has leaked ✓.

ℓ **5/5 marks awarded** This answer just gets 5 marks, although there is a lot of detail missing.

(b) Because this gives more accurate results.

ℓ **0/1 mark awarded** This answer does not provide any specific information relating to this experiment.

(c) (i) Between 0 and 30 the absorbance goes up slightly ✓. Above 40 °C it goes up quickly ✗. Then it starts to level out at about 70 °C ✓.

ℓ **2/3 marks awarded** Student A has correctly identified the three main regions of the graph, stating where changes in gradient occur. However, the term 'quickly' is not correct because the graph does not show anything about time. He/she could have gained a third mark by quoting some figures from the graph.

(ii) High temperatures damage the proteins in the membrane ✓. They get denatured, so they leave holes ✓ in the membrane that the beetroot pigment can get through.

ℓ **2/4 marks awarded** This is a good answer as far as it goes and it is clearly expressed. However, the student should recognise that this is not a sufficiently detailed answer to gain all four available marks.

Student B

(a) Cut lots of pieces of beetroot, all exactly the same size ✓ and all from the same beetroot ✓. Wash the beetroot pieces ✓. Immerse them in the same volume ✓ of water. Put the tubes in water baths at different temperatures, ranging from 0 to 90 °C ✓. After 10 minutes ✓, take the beetroot pieces out of the tubes. Put the water into a colorimeter using a green filter and measure the absorbance of each lot of water. The greater the absorbance, the more the red pigment has escaped so the more permeable ✓ the membrane.

ℓ **5/5 marks awarded** This is an excellent answer for full marks.

(b) Beetroot pigment is red, so it absorbs all other colours of light and reflects red light ✓. If she had used a red filter, she would not have seen any difference between the tubes.

ℓ **1/1 mark awarded** Again, this is a good answer.

> **(c) (i)** The general trend is that the higher the temperature, the greater the absorbance. Between 0 and 30 °C, the absorbance increases slightly ✓ from 16 to 18 arbitrary units. Above 40 °C, it increases much more steeply ✓, levelling off at about 75 °C ✓. The maximum absorbance is 98 arbitrary units.

e **3/3 marks awarded** Student B has given another good answer. However, although the student has quoted some figures from the graph, he/she has not manipulated them in any way — for example, the student could have calculated the increase in absorbance between 0 and 30 °C.

> **(ii)** As temperature increases, the phospholipids and protein molecules in the membrane move about faster ✓ and with more energy. This leaves gaps in the membrane, so the beetroot pigment molecules can get through ✓ and escape from the cell. The protein molecules start to lose their shape at high temperatures ✓ because their hydrogen bonds break ✓, so the protein pores get wider ✓, which increases permeability.

e **4/4 marks awarded** Again, this is an excellent answer that gains full marks.

Question 4

Figure 4 shows the double circulatory system of a mammal.

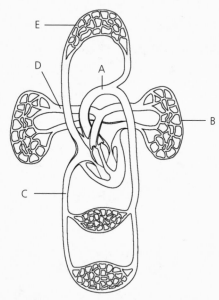

Figure 4

(a) Give the letter of a structure that matches each of the following descriptions.
   (i) An artery containing deoxygenated blood. (1 mark)
   (ii) The place where blood is oxygenated. (1 mark)
   (iii) A vessel that contains valves along its length. (1 mark)
   (iv) The vessel in which the blood is at its highest pressure. (1 mark)

(b) Outline the advantages of a double circulatory system compared with a single circulatory system.

(2 marks)

(c) Describe and explain what happens when the muscles in the walls of the ventricles of the heart contract.

(6 marks)

Total: 12 marks

ⓔ Note that part (c) has two command words — 'describe' and 'explain'.

Student A

(a) (i) D ✓

  (ii) E ✗

  (iii) C ✓

  (iv) A ✓

ⓔ **3/4 marks awarded** (ii) is incorrect — student A has not looked carefully at the diagram and has just assumed that the capillaries at the top are in the lungs.

(b) In a double circulatory system, the blood gets pushed twice as it goes round the body, so it moves faster.

ⓔ **0/2 marks awarded** This is a weak answer. The student has not stated anything incorrect, but the description is not full enough or precise enough to be awarded any marks.

(c) The ventricles get smaller ✓, so the pressure of blood inside them gets bigger ✓. So the blood is pushed up into the aorta and the pulmonary arteries ✓. It cannot go back up into the atria because the valves shut. The pressure in the arteries goes up and down as the heart contracts and relaxes.

ⓔ **3/6 marks awarded** Student A has been given a mark for saying that the ventricles get smaller, but that is quite generous and he/she should really have said that the volume of the ventricles gets smaller. The student has mentioned valves, but doesn't make clear which ones or why they shut. The statement about the pressure going up and down is irrelevant as the question asks only what happens when the muscle contracts.

Student B

(a) (i) D ✓

  (ii) B ✓

  (iii) C ✓

  (iv) A ✓

ⓔ **4/4 marks awarded** All correct for full marks.

(b) In a single circulatory system such as in a fish, the blood leaves the heart at high pressure and then loses pressure as it passes through the capillaries in the gills, where oxygen is absorbed and carbon dioxide lost. The oxygenated blood then goes directly to the tissues, but it is travelling quite slowly and at low pressure. In a double circulatory system such as in a human, the blood returns to the heart after passing through the capillaries in the lungs and the heart increases its pressure ✓ so that it travels to the body tissue rapidly. This provides much more oxygen more effectively to the respiring tissues ✓, enabling a much higher rate of metabolism ✓.

ⓔ **2/2 marks awarded** This is a full and entirely correct answer. Note that no marks are given until the student has made comparative points (e.g. a 'higher rate of metabolism') or has made equivalent statements about both types of system (e.g. what happens in the gills and the lungs). Note that three relevant points are made, but the student gains only 2 marks because this is the maximum score available.

(c) When muscle contracts it gets shorter, so this decreases the volume ✓ inside the ventricle and increases the pressure ✓. Blood is forced upwards against the atrioventricular valves and shuts them ✓ so blood cannot go up into the atria ✓. Blood is also forced upwards against the semilunar valves in the arteries and opens them ✓.

ⓔ **5/6 marks awarded** This is a better answer than student A's response, but it does not get full marks because it does not state which arteries the blood is pushed into.

Question 5

The Irish Threatened Plant Genebank was set up in 1994 with the aim of collecting and storing seeds from Ireland's rare and endangered plant species. The natural habitat of many of these species is under threat.

For each species represented in the bank, seeds are separated into active and base collections. The active collection contains seeds that are available for immediate use, which could be for reintroduction into the wild or for germination to produce new plants and therefore new seeds. The base collection is left untouched. Some of the base collection is kept in Ireland and some is kept at seed banks in other parts of the world.

(a) Suggest why seed banks separate stored seeds into active and base collections. (2 marks)

In 2001, an investigation was carried out into the effect of long-term storage on the ability of the seeds to germinate. Fifteen species were tested. In each case, 100 seeds were tested. It was not possible to use more because in many cases this was the largest number that could be spared from the seed bank. In most cases, the germination rate of the seeds had already been tested when they were first collected in 1994, so a comparison was possible with the germination rates in 2001 after 7 years of storage.

Figure 5 shows the results for two species, *Asparagus officinalis* and *Sanguisorba officinalis*.

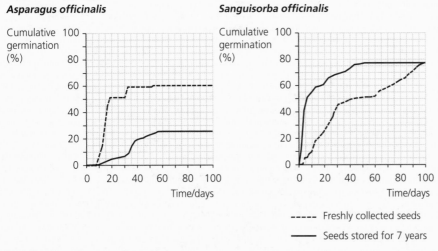

Figure 5

(b) (i) What is the probability that any one freshly collected seed of *Asparagus officinalis* will have germinated after 40 days? (1 mark)

　A 0.06　　　B 0.6　　　C 6.0　　　D 60

(ii) Compare the percentage germination of stored and fresh seeds of *Sanguisorba officinalis*. (3 marks)

(iii) Compare the effect of storage on the percentage germination of *Sanguisorba officinalis* and *Asparagus officinalis*. (3 marks)

(c) It has been suggested that species stored as seeds in seed banks have different selection pressures acting on them compared with the same species living in the wild.

(i) Explain why the selection pressures in a seed bank and in the wild are likely to be different. (2 marks)

(ii) Discuss the possible harmful effects of these differences and suggest how they could be minimised. (5 marks)

Total: 16 marks

ⓔ Here is another question that needs careful reading. Parts (b) (ii) and (iii) both require comparisons, so you need to do more than just describe one and then the other. Use words such as 'more', 'greater', 'but' and 'however'.

Questions & Answers

(a) So they always have some spare.

ⓔ **0/2 marks awarded** The student has not written enough for a mark.

(b) (i) D ✗

ⓔ **0/1 mark awarded** The student has correctly read the value of 60 from the graph, but has failed to appreciate that this is the percentage of seeds that has germinated, not the probability of any one seed germinating. The correct answer is option B, 0.6.

> **(ii)** The stored ones germinated better than the fresh ones. The stored ones go up more quickly than the fresh ones. But they all end up at the same place, about 80% ✓.

ⓔ **1/3 marks awarded** This answer loses out by poor wording and not being clear enough about what exactly is being described. The word 'better' in the first sentence could mean that the seeds germinated more quickly or that more of them germinated, so the student needs to clarify this. 'Go up more quickly' is also not clearly related to germination. The last sentence is quite generously given a mark for the idea that eventually about 80% of the seeds in each batch germinated.

> **(iii)** The *Sanguisorba* seeds germinated better when they had been stored, but the *Asparagus* seeds germinated better when they were fresh. Storing the *Sanguisorba* seeds made them germinate faster, but the *Asparagus* seeds germinated slower ✓.

ⓔ **1/3 marks awarded** Again, poor wording means that this answer gets only 1 mark. The word 'better' is used in the first sentence and, as has been explained above, this is not a good word to use in this context. 1 mark is given for the idea that storage caused slower germination in *Asparagus* but faster germination in *Sanguisorba*.

> **(c) (i)** The conditions in which the seeds grow might be different in the seed bank from in the wild.

ⓔ **0/2 marks awarded** This is not quite correct as seeds do not grow. Growth only happens after germination, so it is seedlings and plants that grow. The student needs to think more carefully about what happens to seeds in a seed bank.

> **(ii)** The seeds could be grown in conditions like those in the wild.

ⓔ **0/5 marks awarded** Again, this answer has not been clearly thought out.

Student B

(a) Having active collections is good because it means there are seeds available that can be used for something. But you must always keep some seeds in storage because the whole point of a seed bank is that it stores seeds and these need to be kept safe so they don't get destroyed ✓. If all of them got used for growing plants, perhaps the plants would die ✓ and you wouldn't have any seeds left.

e **2/2 marks awarded** This is not expressed well but the right ideas are there.

(b) (i) B ✓

e **1/1 mark awarded** This is the correct answer.

(ii) The stored seeds germinated much faster ✓ than the fresh ones. By 10 days, about 60% of the stored seeds had germinated, but only about 8% ✓ of the fresh ones. By 50 days, all of the stored seeds that were going to germinate (about 80%) had germinated. It took 100 days for 80% of the fresh seeds to germinate ✓. We cannot tell if any more would have germinated after that because the line is still going up when the graph stops ✓.

e **3/3 marks awarded** Clear comparative points have been made about the speed at which germination happened and about the maximum percentage of seeds that germinated.

(iii) Storage seemed to help the germination for *Sanguisorba*, but it made it worse for *Asparagus*. For *Asparagus*, storage made it germinate slower, but it germinated faster for *Sanguisorba* ✓. For *Asparagus*, only about 30% of the stored seeds germinated compared with 60% of the fresh seeds ✓, but with *Sanguisorba* about the same percentage of seeds germinated for both fresh and stored ✓.

e **3/3 marks awarded** This is a good answer that makes clear comparisons and discusses both the speed of germination and the percentage of seeds that eventually germinated.

(c) (i) In the seed bank, the seeds are just stored. So the ones that survive are the ones that are best at surviving in those conditions as dormant seeds ✓. In the wild, the plants have to be adapted to grow in their habitat ✓, so maybe they have to have long roots or big leaves or whatever. In the seed bank, that doesn't matter. So you might get seeds that are really good at surviving in a seed bank, but when they germinate they produce plants that aren't good at surviving in the wild.

e **2/2 marks awarded** This is a good answer that answers the question but it could have been written a little more carefully and kept shorter.

> **(ii)** You could keep on collecting fresh seeds from plants in the wild ✓, and only storing them for a little while before replacing them with new ones ✓. If you had to store seeds for a long time, you could keep germinating some of them ✓ and growing them in conditions like in the wild ✓ and then collect fresh seeds from the ones that grew best ✓.

ⓔ 5/5 marks awarded This is a good answer to a tricky question. The student has made several sensible suggestions, including storing the seeds for a shorter time and periodically exposing the plants to natural conditions where the 'normal' selection pressures will operate.

Question 6

A student investigated the effect of wind speed on the rate of transpiration of a plant. The soil in a pot containing a plant was watered and then covered with plastic. The pot was placed on a top pan balance and a fan placed nearby to cause air movement around the plant's leaves.

The mass of the plant was measured at 5-minute intervals for 30 minutes. The fan was then switched off and the mass measured for a further 30 minutes. The results are shown in Figure 6.

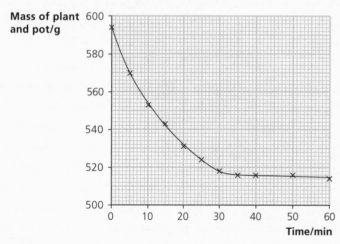

Figure 6

(a) (i) Draw a tangent to the line at 10 minutes and use your tangent to calculate the rate of loss of mass at this time.　　(3 marks)

(ii) Explain why there is a change in the gradient of the graph just after 30 minutes.　　(3 marks)

(b) Water moves across the root of a plant through both apoplastic and symplastic pathways. Explain the difference between these two pathways.　　(3 marks)

(c) Water moves up the stem of a plant and into the leaves through xylem vessels. Explain how the structure of xylem vessels is related to their role in the transport of water.　　(4 marks)

Total: 13 marks

ⓔ Draw the tangent in part (a) (i) carefully and show clearly how you arrive at your answer. Part (b) requires a comparison; try to do this, rather than just describing one pathway and then the other.

Student A

(a) (i)

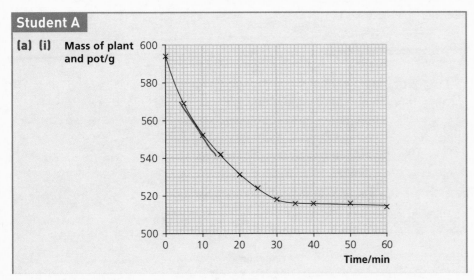

ⓔ **0/3 marks awarded** The student has attempted to draw a tangent to the curve, but this is not precisely at 10 minutes, so does not earn a mark. There is no attempt to calculate the rate of loss of mass.

(ii) This is when the fan was switched off. Before that, there was wind blowing on the plant's leaves, so it was transpiring faster ✓. When the fan was switched off, transpiration slowed down, so the rate of water loss was less ✓.

ⓔ **2/3 marks awarded** This answer gets 1 mark for identifying that the rate of water loss decreased when the fan was switched off and another mark for explaining that this happened because of a change in the rate of transpiration. A more detailed explanation is required for the third mark.

(b) Apoplastic is between the cells and symplastic is through the cells ✓. Water mostly moves by the apoplastic pathway.

ⓔ **1/3 marks awarded** The answer is lacking in detail, so only 1 mark is awarded.

(c) Xylem vessels are made of dead cells, so they are hollow which leaves a space that water can move through without any cytoplasm getting in the way ✓. They don't have any end walls, so they join up with the next xylem vessel and the water can go through from one to the next easily ✓. They have strong walls containing lignin, which stops them caving in when the pressure inside them gets low ✓.

ⓔ 3/4 marks awarded This is a good answer in which the student has linked the structure of the xylem vessel to particular ways in which it is adapted for water transport. However, only three distinct points have been made.

Student B

(a) (i)

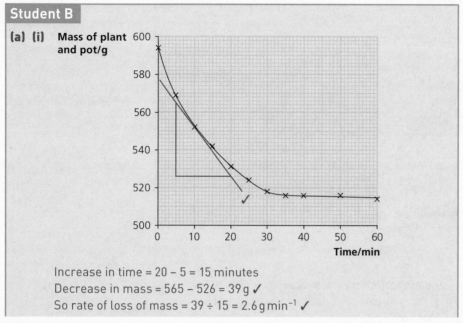

Increase in time = 20 − 5 = 15 minutes
Decrease in mass = 565 − 526 = 39 g ✓
So rate of loss of mass = 39 ÷ 15 = 2.6 g min^{-1} ✓

ⓔ 3/3 marks awarded The student's tangent has been carefully drawn, and a mark is awarded for that. The lengths of the vertical and horizontal sides of the triangle have been correctly recorded, and the calculation is correct, giving a final answer with the right units.

(ii) The fan was switched off at 30 minutes, so wind speed suddenly dropped. High wind speed increases the rate of transpiration ✓ because it moves away the humid air that otherwise collects next to the leaf surface ✓. With no fan, the water vapour that diffused out through the stomata remained next to the surface and this decreased the diffusion gradient ✓, so now water diffused out more slowly. The loss in mass of the plant was caused by the loss of water so, if transpiration was slower, the loss in mass was slower too ✓.

ⓔ 3/3 marks awarded This is an excellent answer, expressed clearly and providing a full explanation in terms of transpiration, the effect of wind and diffusion gradients.

(b) Most water moves across the root by the apoplastic pathway. This is made up of the cell walls of the plant cells and also the spaces between them ✓. Water can seep across this pathway without ever going through a cell. Water can also travel by the symplastic pathway, which involves crossing cell membranes and going into cells ✓. It has to move across the cell membranes by osmosis ✓ and this is a slower way of travelling than through the apoplast.

ⓔ **3/3 marks awarded** This is a good answer that gains all 3 marks.

(c) Xylem vessels are narrow, dead, empty cells with lignified walls. Being narrow helps the water column inside them to be supported and not break ✓. Because they are empty, there is nothing in the way of the water as it moves up by mass flow ✓, pulled by transpiration pull (cohesion-tension theory). The lignin helps to hold them firm so that when there is low pressure inside them ✓, caused by transpiration pull, they don't collapse. Many xylem vessels are joined end to end making a continuous pathway ✓ from the roots right up into all the leaves.

ⓔ **4/4 marks awarded** This is a better answer than student A's response. The mention of the cohesion–tension theory does not gain credit because there is no explanation of what it is. Nevertheless, this is not necessary for full marks.

■Sample paper 2

Question 1

Tigers, *Panthera tigris*, are a threatened species in the wild, largely due to loss of habitat and also hunting.

Tigers living in different parts of the world have measurable phenotypic differences from each other and have been assigned to different subspecies. For example, the Bengal tiger, *P. tigris tigris*, lives in India, whereas the Amur tiger, *P. tigris altaica*, lives in Siberia. Amur tigers tend to have thicker fur with paler golden stripes than Bengal tigers, and they are heavier.

Zoos have been breeding tigers in captivity for over a century and these captive tigers may be all that can save the species from extinction. For example, in 2008 there were 420 Amur tigers in captive breeding programmes and about the same number living wild in Siberia. One subspecies, *P. tigris amoyensis*, the South China tiger, now exists only in captivity.

In the past, breeding programmes allowed tigers from different subspecies to breed with one another, but today care is taken that captive tigers breed only with their own subspecies in order to maintain distinct populations of each. Nevertheless, attempts are made to ensure that each tiger breeds only with one to which it is not closely related as this helps to maintain genetic diversity. This has been successful, and a survey carried out in 2008 showed that genetic diversity among the captive population of Amur tigers was significantly greater than the genetic diversity among wild Amur tigers.

(a) Explain how natural selection could account for the differences between Amur tigers and Bengal tigers. (5 marks)

(b) (i) Explain the meaning of the term 'genetic diversity'. (2 marks)

   (ii) Suggest why the captive population of Amur tigers has a greater genetic diversity than the wild population. (3 marks)

   (iii) Explain why it is desirable for a population to have a reasonably high genetic diversity. (3 marks)

Total: 13 marks

ⓔ Take time to read the passage carefully before you try to answer the questions, even if you feel under time pressure.

Student A

(a) It is very cold in Siberia, so it is an advantage for the tigers to have thick fur ✓. The lighter coloured stripes could help them to be camouflaged against the snow.

ⓔ **1/5 marks awarded** The student has correctly identified a feature of the Amur tigers that adapts them for life in a cold climate, but he/she has not explained anything at all about natural selection.

(b) (i) This means all the different alleles of the genes ✓.

ⓔ 1/2 marks awarded This is a correct description, but the student needs to go a little further to get the second mark.

(ii) Maybe the tigers in the wild are all closely related ✓ and they all just interbreed with their own family ✓. The tigers in captivity have been bred with other unrelated ones ✓.

ⓔ 3/3 marks awarded Although this is not a strong answer, it does just manage to make three good points and so gets full marks.

(iii) The more different genes there are, the more the tigers will vary. If the climate changes ✓, some of them might be able to survive, or if there is a new disease.

ⓔ 1/3 marks awarded Student B shows that he/she understands that the value of genetic diversity is the ability to survive in different or changing environments, by means of an example. To get more marks, this idea needs to be developed further.

Student B

(a) Natural selection means that only the best-adapted animals survive. Amur tigers with genes for thick fur are better insulated ✓, so they keep warmer and are more likely to survive ✓ in a cold climate than tigers with thin fur. So the thick-furred tigers are more likely to reproduce ✓ and pass on their thick-fur alleles ✓ to their offspring. This goes on happening for many generations ✓, until all the tigers have thick fur.

ⓔ 5/5 marks awarded This is a clear explanation. Another point that could have been made near the beginning of the answer is that there will be natural variation in the tiger population, with some tigers having thicker fur than others.

(b) (i) All the different alleles ✓ in the gene pool ✓ of a population.

ⓔ 2/2 marks awarded This is correct for both marks.

(ii) The tigers in zoos have been part of a captive breeding programme, where everyone makes sure that they breed only with tigers that they are not related to ✓. In the wild, tigers have large territories so they don't meet each other very often ✓. So the tigers that mate together may all be descended from just a few ✓ older tigers that always lived there and so they all have similar alleles.

ⓔ 3/3 marks awarded This response has the right ideas. The student has thought about how tigers probably live in the wild and surmised that there is less chance of a tiger meeting and mating with an unrelated animal than there is in zoos.

> **(iii)** This is so if there is a new disease ✓ then maybe at least some of them will be immune to it and they won't all die ✓.

ⓔ 2/3 marks awarded This is correct, but there is not quite enough for full marks.

Question 2

Figure 1 shows the fluid mosaic model of membrane structure.

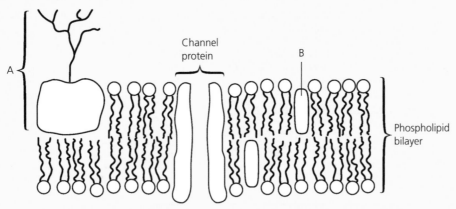

Figure 1

(a) Name molecules A and B. (2 marks)

(b) Explain how the properties of phospholipids cause them to form a bilayer. (3 marks)

(c) Explain why the representation of membrane structure is said to be a 'model'. (2 marks)

(d) Sodium ions can pass across cell membranes by facilitated diffusion or by active transport.

    (i) Explain why it is not possible for sodium ions to diffuse freely through the phospholipid bilayer. (2 marks)

    (ii) With reference to Figure 1, explain how sodium ions could move across the membrane by facilitated diffusion. (2 marks)

Table 1 shows the concentrations of three substances in the cytoplasm and in the solution outside the cell.

Substance	Concentration inside cell/$mol\,dm^{-3}$	Concentration outside cell/$mol\,dm^{-3}$
P	0.23	0.24
Q	0.02	0.13
R	0.10	0.04

Table 1

(e) **(i)** Calculate the percentage difference in the concentration of substance Q on the two sides of the cell-surface membrane. Show your working. (3 marks)

    (ii) By which process or processes in Table 2 could substances P and R have moved across the membrane? (1 mark)

	P	R
A	Facilitated diffusion or active transport	Diffusion or facilitated diffusion
B	Diffusion or facilitated diffusion	Active transport only
C	Diffusion only	Facilitated diffusion or active transport
D	Active transport only	Facilitated diffusion only

Table 2

Total: 15 marks

e Several parts of this question ask you to 'explain', so you need to say how or why.

Student A

(a) A = protein ✓, B = cholesterol ✓

e **2/2 marks awarded** Both answers are correct for 2 marks.

(b) They have hydrophobic heads that go towards water and hydrophilic tails that go away from it ✓.

e **1/3 marks awarded** Student A has muddled the terms 'hydrophobic' and 'hydrophilic'. However, he/she has correctly stated that the heads and tails 'go' towards and away from water respectively, which is not well expressed but is just enough for 1 mark.

(c) It is just something that people have made up, not the real thing. We cannot actually see a membrane looking like this ✓.

e **1/2 marks awarded** This is true, but student A fails to explain that the structure of the membrane has not just been 'made up' — it has been deduced from experimental evidence about how membranes behave.

(d) (i) They have a charge ✓ on them, so they cannot go through the phospholipids which are hydrophobic.

e **1/2 marks awarded** Student A has the right idea, but needs to make clear that it is the phospholipid tails that are hydrophobic.

(ii) They could go through the protein channel ✓, down their concentration gradient ✓.

e **2/2 marks awarded** This is concise and entirely correct.

(e) (i) Difference = 0.13 − 0.02 = 0.11 ✓.

So percentage difference $= \dfrac{0.11}{0.13} = 0.85$ ✓

ℰ 2/3 marks awarded The calculation is correct, except that student A has forgotten to multiply by 100 to find the percentage.

> **(ii)** C ✗

ℰ 0/1 mark awarded This is incorrect. Student A may not appreciate that facilitated diffusion happens down a diffusion gradient and therefore results in equal concentrations on each side of a membrane.

Student B

(a) A = protein ✓, B = cholesterol ✓

ℰ 2/2 marks awarded Both answers are correct for 2 marks.

> **(b)** Their hydrophilic heads are attracted to water ✓ and their hydrophobic tails try to get away from it ✓, which they can do by putting their tails together and their heads facing outwards ✓.

ℰ 3/3 marks awarded A clear and correct answer.

> **(c)** The structure of the membrane has been worked out because we know a lot about the properties of membranes and this model can explain all of those properties ✓. So even though we cannot see ✓ all the molecules in a membrane, we are fairly sure this is what it would be like.

ℰ 2/2 marks awarded This is a good answer.

> **(d) (i)** Sodium ions are only small, but they have a positive charge ✓ so they are repelled from the hydrophobic tails ✓ of the phospholipids and cannot get through.

ℰ 2/2 marks awarded This is correct for both marks.

> **(ii)** They can diffuse through the protein channels ✓.

ℰ 1/2 marks awarded This is correct, but the answer does not explain what 'diffuse' means so has only given half of the required explanation.

> **(e) (i)** Concentration inside – concentration outside = 0.11 ✓
> Concentration inside = 0.02
> So percentage difference = $\dfrac{0.11}{0.02} \times 100$ ✓ = 550%
> greater outside than inside ✓

ℰ 3/3 marks awarded This is all correct. This student has calculated the difference as a percentage of the concentration inside the cell, whereas student A calculated it compared to the concentration outside. Both are acceptable methods.

(ii) B ✓

ⓔ 1/1 mark awarded This is correct. Substance P has the same concentration on both sides of the membrane, so could have moved by diffusion or facilitated diffusion. Substance R has different concentrations, so must have moved by active transport.

Question 3

(a) Describe how gel electrophoresis can be used to separate DNA fragments of different lengths. (6 marks)

Figure 2 shows an electrophoresis gel that was made using DNA samples from four species of deer.

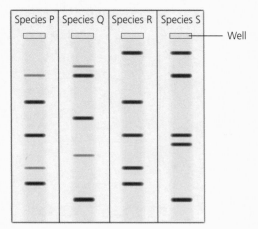 ──── Well

Figure 2

(b) (i) Which two species appear to be most closely related? (1 mark)

    A P and Q   B P and R   C R and S   D Q and S

(ii) The binomial of species P is *Odocoileus virginianus*. Complete the table to show the classification of this species. (4 marks)

Domain	
Kingdom	
	Chordata
	Artiodactyla
Family	Cervidae
Genus	
Species	

Total: 11 marks

ⓔ Take care to write enough in part (a) to give you a good chance of getting all 6 marks.

Questions & Answers

a) First, you cut the DNA up into pieces using restriction enzymes ✓. Then you put the DNA on to some agarose gel ✓ in a tank and switch on the power supply so the DNA gets pulled along the gel ✓. The bigger pieces move more slowly, so they end up not so far along the gel as the smaller pieces ✓. You cannot see the DNA, so you need to stain it with something so it shows up.

ⓔ **4/6 marks awarded** There are no errors in this answer, but a little more detail is needed.

(b) (i) B ✓

ⓔ **1/1 mark awarded** This is correct. There are four shared bands between P and R, more than any of the other pairs.

(ii)

Domain	Eukaryotes ✓
Kingdom	Animalia ✓
Class ✗	Chordata
Phylum ✗	Mammalia ✓
Order ✓	Artiodactyla
Family	Cervidae
Genus	*Odocoileus* ✓
Species	*virginianus* ✓

ⓔ **3/4 marks awarded** The student has confused the sequence of class and phylum, but otherwise everything is correct.

(a) The DNA is cut into fragments using restriction enzymes ✓, which cut it at particular base sequences. Then you place samples of the DNA into little wells in agarose gel ✓ in an electrophoresis tank. A voltage is then applied ✓ across the gel. The DNA pieces have a small negative charge so they steadily move towards the positive ✓ terminal. The larger they are, the more slowly they move ✓, so the smaller ones travel further ✓ than the big ones. After a time, the power is switched off so the DNA stops moving. You can tell where it is by using radioactivity ✓, so the DNA shows up as bands on a photographic film.

ⓔ **6/6 marks awarded** This is all correct and there is enough detail for full marks.

(b) (i) B ✓

ⓔ **1/1 mark awarded** This is correct.

(ii)

Domain	Eukaryotes ✓
Kingdom	Animalia ✓
Phylum ✓	Chordata
Class ✓	Mammalia ✓
Order ✓	Artiodactyla
Family	Cervidae
Genus	*Odocoileus* ✓
Species	*virginianus* ✓

ⓔ 4/4 marks awarded All correct for full marks.

Question 4

Figure 3 shows pressure changes in the left atrium and left ventricle of the heart and the aorta during the cardiac cycle.

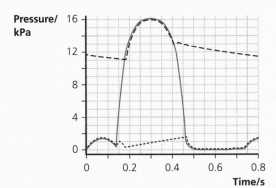

Key

-- Aorta

— Left ventricle

···· Left atrium

Figure 3

(a) (i) Calculate how many heart beats there will be in 1 minute. (2 marks)

(ii) Calculate the mean rate of change of pressure in the left ventricle between 0.3 seconds and 0.5 seconds. (2 marks)

(b) (i) On Figure 3, indicate the point at which the semilunar valves in the aorta snap shut. (1 mark)

(ii) Explain what causes the semilunar valves to shut at this point in the cardiac cycle. (2 marks)

(iii) On Figure 3, indicate the period when the left ventricle is contracting. (1 mark)

(iv) On Figure 3, draw a line to show the changes in pressure in the right ventricle. (2 marks)

(c) After the blood leaves the heart, it passes into the arteries. The blood pressure gradually reduces and becomes more steady as the blood passes through the arteries. Explain what causes this reduction and steadying of the blood pressure. (3 marks)

Total: 13 marks

(e) Most of this question is about being able to interpret a graph precisely. Take care to be as accurate as possible with your answers to parts (b) (i), (iii) and (iv).

Student A

(a) (i) 1 cycle in 0.8 seconds ✓ so in 60 seconds there will be 60 × 0.8 ✗
= 48 beats

(e) **1/2 marks awarded** The student has read the length of one cycle correctly but the calculation is wrong.

(ii) It drops from 16 to 0 kPa in 0.2 s ✓, so the mean rate of change is:
$$\frac{16}{0.2} = 80 \, kPa \, ✗$$

(e) **1/2 marks awarded** The student has calculated the mean rate of change correctly, but has not given correct units.

(b)

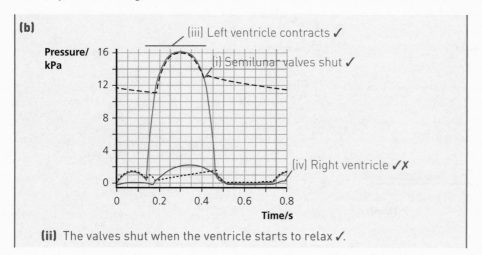

(iii) Left ventricle contracts ✓

(i) Semilunar valves shut ✓

(iv) Right ventricle ✓✗

(ii) The valves shut when the ventricle starts to relax ✓.

(e) **4/6 marks awarded** The annotations for parts (i) and (iii) are correct. The answer to (ii) is correct as far as it goes, but student A needs to give more information in order to get the second mark. The annotation for part (iv) is partly correct. The pressure in the right ventricle is correctly shown as less than that in the left ventricle, but it should be contracting and relaxing at exactly the same times as the left ventricle.

(c) The pressure gets less as the blood gets further away from the heart ✓. The muscle in the walls of the arteries contracts ✗ and relaxes to push the blood along and it does this in between heart beats so the pulse gets evened out.

(e) **1/3 marks awarded** The first statement is correct. However, it is not correct that the muscles in the artery wall contract and relax.

Student B

(a) (i) $\frac{60}{0.8}$ = 75 beats per minute ✓✓

e **2/2 marks awarded** This is entirely correct for both marks.

(ii) Drop in pressure is 16 kPa over a time period of 0.2 s ✓.
Therefore, the mean rate of change =
$\frac{16}{0.2}$ = 80 kPa per s, or 8 kPa in 0.1 s ✓

e **2/2 marks awarded** This is correct for both marks.

(b)

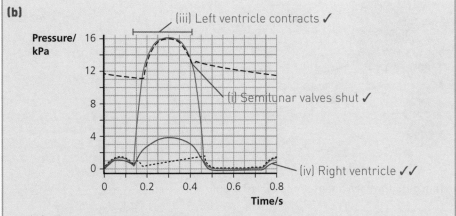

(ii) They close when the pressure of the blood inside the arteries is higher than inside the ventricles ✓. The blood therefore pushes down on the valves and makes them shut ✓.

e **6/6 marks awarded** These are all correct for full marks.

(c) As the blood is forced into the artery as the ventricle contracts ✓, it pushes outwards on the artery wall, making the elastic tissue stretch ✓. In between heart beats, the pressure of the blood inside the artery falls and the elastic tissue recoils ✓. So the wall keeps expanding and springing back. When it springs back, it pushes on the blood in between ✓ heart beats, so this levels up the pressure changes.

e **3/3 marks awarded** This answer explains well why the blood pressure levels out. However, it does not mention the overall fall in blood pressure. All the same, this is an excellent answer that easily gets full marks.

Question 5

Figure 4 shows the structure of part of a gill of a fish.

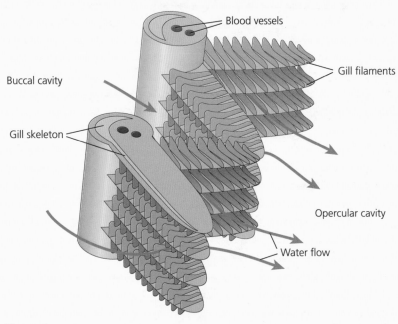

Figure 4

(a) List three features of the fish gills that are adaptations for efficient
gas exchange. (3 marks)

Most fish have haemoglobin in their blood, which helps to transport oxygen.
Some fish, such as tuna, also have large quantities of myoglobin in their muscles,
which makes their flesh looks red.

Figure 5 shows an oxygen dissociation curve for haemoglobin in skipjack tuna.

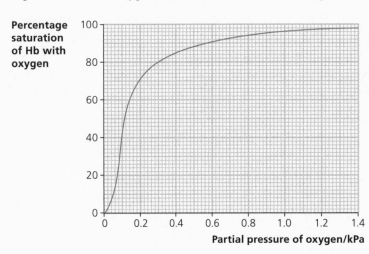

Figure 5

(b) (i) Use Figure 5 to explain how haemoglobin transports oxygen from the gills of a tuna to its respiring tissues. You should include reference to the structure of the haemoglobin molecule in your answer. (6 marks)

(ii) The myoglobin in the muscles of skipjack tuna acts as an oxygen store. On Figure 5, sketch a line to show the oxygen dissociation curve for myoglobin. (2 marks)

Total: 11 marks

🅔 Part (b) requires you to bring together your own knowledge, as well as the information supplied in Figure 5, to construct an explanation. This is not easy, and is worth 6 marks, so take time to plan your answer before you begin to write.

Student A

(a) They have a big surface area because of all the lamellae ✓. They are really thin ✓ so it isn't far from the water to the blood. The blood and the water flow in opposite directions ✓, which makes is easier for the oxygen to go from the water into the blood.

🅔 **3/3 marks awarded** Three good points are made. Note that the question only asked for the features to be listed, so there was no need to give explanations.

(b) (i) Haemoglobin is a protein that has four haem groups ✓ that can combine with oxygen when there is a high concentration. You can see from Figure 5 that when the pressure of oxygen is 1.4 kPa almost 100% of the haemoglobin is combined with it ✓. This is what happens in the gills ✓. As the blood flows to the tissues, the haemoglobin gradually loses its oxygen until it is 0% saturated, which means it has given up all its oxygen. When it is combined with oxygen, it is called oxyhaemoglobin and is red, but when it loses its oxygen it is blue.

🅔 **3/6 marks awarded** Student A begins well, with a correct statement about the structure of haemoglobin and how it combines with oxygen, as well as relating a high saturation to a particular partial pressure of oxygen and stating that this is the situation in the gills. However, it is not correct that the haemoglobin 'gradually loses its oxygen' as it travels to the tissues, nor is it likely that it becomes '0% saturated'. The information about the colour of oxyhaemoglobin and haemoglobin is irrelevant.

(ii)

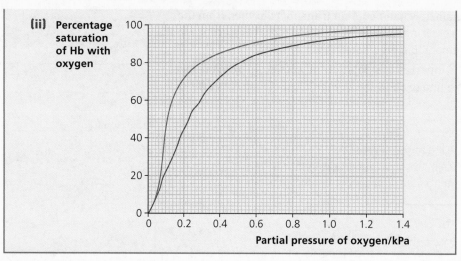

0/2 marks awarded This is completely incorrect. The myoglobin curve should lie to the left of the haemoglobin curve and reach saturation at a lower partial pressure of oxygen.

Student B

(a) (1) Very thin ✓. (2) Counter-current flow (blood and water flow in opposite directions) ✓. (3) Large surface area ✓.

ⓔ **3/3 marks awarded** These are all correct and it is an appropriately concise answer to a 'list' question.

(b) (i) The dissociation curve indicates that haemoglobin is good at picking up oxygen in the gills ✓, where the partial pressure of oxygen is high, and releasing it in the tissues ✓, where respiration is happening and so partial pressure of oxygen is much lower ✓. We can see this because Figure 5 shows that there is a high percentage saturation of haemoglobin when the oxygen partial pressure is high and a much lower saturation when the oxygen partial pressure is low ✓, showing that the haemoglobin releases its oxygen when it gets to the tissues. Haemoglobin is a globular protein with a quaternary structure ✓, made of four polypeptide chains each with a haem group. When one of the haem groups combines with oxygen ✓, it changes the shape of the molecule so it is easier for the other haem groups to bind ✓.

ⓔ **6/6 marks awarded** This is a full and clear answer that refers to Figure 5 as asked in the question. The student refers to the ability of haemoglobin both to pick up and to release oxygen — an essential feature of an oxygen transport molecule. He/she also relates this to the partial pressures of oxygen as shown on the graph and then to the situation in the gills and tissues of the fish. The description of the effect of binding with oxygen on the shape of the haemoglobin molecule is also a relevant point to make.

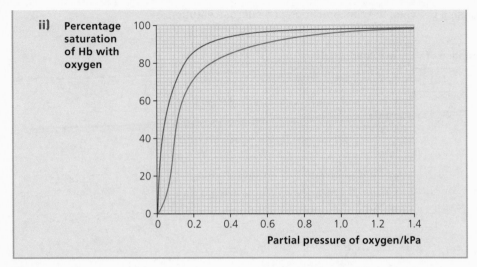

ii) Percentage saturation of Hb with oxygen

e 2/2 marks awarded The curve is correctly drawn to the left of the haemoglobin curve, showing that myoglobin becomes saturated with oxygen at quite low partial pressures and does not give up its oxygen until these partial pressures are very low.

Question 6

A student wanted to find the solute potential of the cell sap in the cells of a red onion.

1   He peeled pieces of epidermis from inside an onion bulb, cut them into equal sized squares and placed them in sucrose solutions of different concentrations. He left them for 20 minutes to give time for equilibrium to be reached.

2   He then viewed each piece of epidermis under a microscope. He counted all the cells in the field of view of one piece of epidermis and recorded how many were plasmolysed.

3   He then moved the slide to count all the cells in a second area of the same piece and then counted how many of these cells were plasmolysed.

4   He then repeated steps 2 and 3 for each of the sucrose concentrations.

His results are shown in Table 3.

Concentration of sucrose solution/ $mol\,dm^{-3}$	0.0	0.2	0.4	0.6	0.8
Solute potential of sucrose solution/kPa					
Number of cells counted	48	56	51	50	47
Number of plasmolysed cells	0	1	18	41	47
Percentage of plasmolysed cells	0.0	1.8	35.3		100.0

Table 3

Figure 6 shows the relationship between the concentration of a sucrose solution and its solute potential.

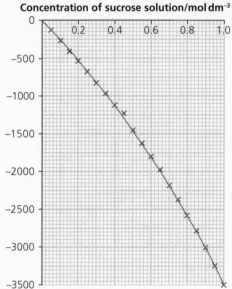

Figure 6

(a) (i) Use Figure 6 to complete the second row of Table 3. (2 marks)

    (ii) Calculate the percentage of plasmolysed cells in the 0.6 mol dm^{-3} sucrose solution. (2 marks)

    (iii) On a graph grid, plot a suitable graph to display these results. (5 marks)

    (iv) When exactly 50% of cells are plasmolysed, we can assume that, on average, the cells are just on the point of becoming plasmolysed (incipient plasmolysis) and therefore their pressure potential is 0.

        Use the following equation and your graph to estimate the water potential of the onion cells. (2 marks)

        water potential = solute potential + pressure potential

Sucrose is transported through a flowering plant in phloem. Figure 7 shows some of the cells that make up the phloem tissue in a plant stem.

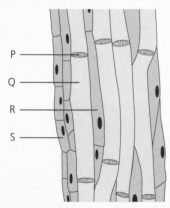

Figure 7

(b) (i) Which labelled structure is a companion cell? (1 mark)

(ii) With reference to Figure 7, explain the mass flow hypothesis for the movement of sucrose through phloem tissue. (5 marks)

Total: 17 marks

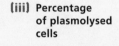 Much of this question involves handling data, in a table, a graph and an equation. Show your working for parts (a) (ii) and (iv). For part (b) (ii), notice that you should refer to Figure 7 in your explanation.

Student A

(a) (i)

Concentration of sucrose solution/mol dm^{-3}	0.0	0.2	0.4	0.6	0.8
Solute potential of sucrose solution/kPa	0.0	540	1120	1800	2580
Number of cells counted	48	56	51	50	47
Number of plasmolysed cells	0	1	18	41	47
Percentage of plasmolysed cells	0.0	1.8	35.3		100.0

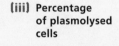 **1/2 marks awarded** The student has read the graph correctly, but has not included the minus signs.

(ii) 82 ✓

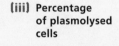 **1/2 marks awarded** The student has done the calculation correctly, but has not given the answer to the same number of decimal places as the other percentages in the table.

(iii)

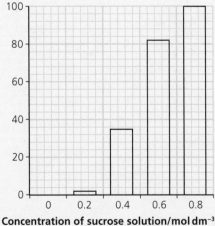

Percentage of plasmolysed cells (y-axis) vs Concentration of sucrose solution/mol dm^{-3} (x-axis)

e 3/5 marks awarded The student has chosen to plot percentage of plasmolysed cells against concentration, which is appropriate. He/she has correctly placed the independent variable (concentration) on the x-axis and the dependent variable on the y-axis. He/she has labelled both axes fully and chosen appropriate scales. However, this should be a line graph, not a bar chart, because both the x-axis and y-axis parameters are continuous variables.

> **(iv)** When 50% of the cells are plasmolysed, water potential = solute potential. So the water potential is somewhere between 0.4 and 0.6, perhaps 0.5.

e 0/2 marks awarded The student has been unable to determine the concentration of sucrose in which 50% of cells are plasmolysed because he/she has drawn the wrong type of graph. He/she has correctly stated that water potential = solute potential in these circumstances, but has not been able to substitute into the equation. Moreover, the values stated are concentrations, not solute potentials.

> **(b) (i)** S ✗

e 0/1 mark awarded S cannot be a companion cell because it is not in contact with a sieve tube.

> **(ii)** The mass flow hypothesis states that sucrose is moved through phloem by mass flow of liquid through the phloem tubes, from high pressure in a source to low pressure in a sink ✓. Sucrose is moved into the phloem by active transport ✓ and water follows by osmosis. This creates a high pressure ✓. At the sink, sucrose is used up, so water leaves by osmosis and creates a low pressure. The sieve plates don't really help with any of this. There isn't much cytoplasm inside the phloem tubes, so there is space for the sucrose solution to flow through ✓.

e 4/5 marks awarded Although this is not a strong answer, there is just enough for 4 marks. The student correctly summarises the mass flow hypothesis and attempts to explain how the pressure gradient is produced, as well as referring to how the structure of the sieve tube elements allows this process to take place. However, there is no mention of companion cells.

Student B

(a) (i)

Concentration of sucrose solution/ mol dm^{-3}	0.0	0.2	0.4	0.6	0.8
Solute potential of sucrose solution/kPa	0.0	−540	−1120	−1800	−2580
Number of cells counted	48	56	51	50	47
Number of plasmolysed cells	0	1	18	41	47
Percentage of plasmolysed cells	0.0	1.8	35.3		100.0

🅔 **2/2 marks awarded** The figures given are entirely correct.

(ii) 82.0 ✓

🅔 **2/2 marks awarded** This is correct for both marks.

(iii)

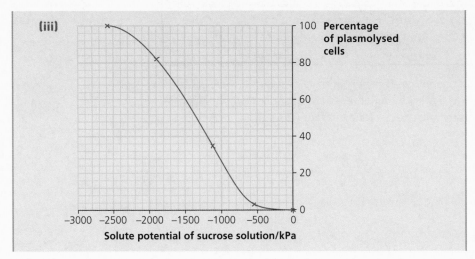

Solute potential of sucrose solution/kPa

🅔 **5/5 marks awarded** This is an excellent graph. The student has chosen to take solute potential as the independent variable, which will help when answering part (iv). The axes are correctly orientated and fully labelled, with suitable scales. The points have been carefully plotted and a good best-fit line has been drawn.

(iv) From my graph, the solute potential of the sucrose solution in which 50% of the onion cells are plasmolysed is −1350 kPa ✓. So at this concentration, turgor pressure = 0 and water potential = solute potential ✓. So we can say the water potential of the onion cells is −1350 kPa ✓.

🅔 **2/2 marks awarded** The correct value has been read from the graph and the equation provided has been used correctly. The answer is given with the correct units.

(b) (i) R ✓

ⓔ **1/1 mark awarded** This is correct for the mark.

(ii) Sucrose is loaded into the phloem by the companion cells using active transport ✓. This decreases the water potential inside the phloem sieve tube element, so water follows because it moves by osmosis down a water potential gradient ✓. This means there is more liquid inside the phloem, so it has a high hydrostatic pressure ✓. This happens in a source, e.g. a leaf that has been photosynthesising ✓.

At the other end of the phloem, sucrose moves out because the cells are using it in a sink, e.g. in the root ✓. So water follows by osmosis and there is less liquid inside the phloem and a lower hydrostatic pressure ✓. So the sucrose solution flows from the source to the sink down a hydrostatic pressure gradient by mass flow ✓.

You can see from Figure 7 that the companion cells are next to the phloem cells so they can do the active loading of sucrose ✓. The phloem cells do not have many contents, making is easy for the sucrose to flow through ✓. Some people think that the mass flow theory cannot be right because of the sieve plates, but others think that these are there to stop the sucrose from running out of the plant if the phloem gets damaged ✓.

ⓔ **5/5 marks awarded** This is a full and entirely correct answer. However, it is rather too long. There are only 5 marks available, so the quantity of detail given is too great. Nevertheless, this is much better than student A's answer.

Knowledge check answers

1

Domain	Eukarya
Kingdom	Animalia
Phylum	Chordata
Class	Reptilia
Order	Squamata
Family	Elapidae
Genus	*Ophiophagus*
Species	*hannah*

2 We only have their structures to compare — and often even these can be only partially reconstructed. We know nothing about their physiology, behaviour or breeding habits. We cannot know if they could reproduce together to produce fertile offspring.

3 a The position of the enzyme on the electrophoresis gel.

b The different forms of the enzyme may have different masses or different R groups that give them different overall charges.

c The enzymes in lanes 1, 2 and 3 could belong to one species because they all have dark areas at approximately the same level, although there is an extra bar in lane 2. Lane 4 has no dark area at this level, so is probably a different species. Lane 5 has a dark area at a different level from any of the others.

4 a

	Human	Chimpanzee	Rhesus monkey	Mouse	Chicken	Clawed toad
Chimpanzee	1					
Rhesus monkey	1	2				
Mouse	5	6	5			
Chicken	10	9	8	9		
Clawed toad	10	11	9	8	7	
Rainbow trout	16	15	15	11	11	13

b Human and chimpanzee, or human and rhesus monkey.

c Human and rainbow trout.

5 The table will depend on your choice of organisms. It should clearly explain what each adaptation is and how it relates to the organism's way of life.

6

Species	n	$n - 1$	$n(n - 1)$
A	2	1	2
B	35	34	1190
C	1	0	0
D	81	80	6480
E	63	62	3906
F	2	1	2
G	5	4	20
H	11	10	110
I	1	0	0
	$N = 201$		$\sum n(n - 1) = 11710$

$$D = \frac{N(N - 1)}{\sum n(n - 1)}$$

$$= \frac{201 \times 200}{11710}$$

$$= 3.43$$

7 Both involve the passive movement of substances down a concentration gradient. However, facilitated diffusion can only occur through protein channels in the cell membrane.

8 It will increase it — that is, make it less negative.

Knowledge check answers

9 They both require energy input from the cell. In endocytosis, this is needed to put out the extensions and to cause the membrane to fuse to form a vacuole. In exocytosis, energy is needed to surround the object with membrane and to move it to the cell membrane.

10 The x-axis of your graph should be labelled 'Temperature/°C' and have a scale of equal intervals running from 0 to 100. The y-axis should be labelled 'Mean absorbance'. You don't know the units or the range here, so you could just put an upward-pointing arrow on this axis. The curve should start at about 0 and go upwards, levelling off at a fairly high temperature — this is because beyond a certain temperature the membrane will already be as leaky as it is going to get. You cannot know exactly what this temperature will be, but 40 °C would be a good guess.

11 Active transport involves the movement of substances up their concentration gradient using energy from the hydrolysis of ATP. Facilitated diffusion involves the movement of substances down their concentration gradient with no energy input from the cell.

12 Time taken for one complete heart beat is about 0.75 seconds. (Look for the time from one particular part of the first heart beat to the time for the same stage of the second one.) Therefore, in 1 minute there will be:

$$\frac{60}{0.75} = 80 \text{ beats}$$

13 All arteries carry blood away from the heart, but not all of them carry oxygenated blood. The pulmonary artery carries deoxygenated blood from the heart to the lungs.

14 The left side of the heart has more muscle in its wall. The contraction of this muscle produces a greater pressure than in the right side.

15 Prothrombin, thrombin, fibrinogen and fibrin.

16 An example of a benefit: if you know you are at high risk of cardiovascular disease, you can take action to reduce other risk factors. An example of a potential problem: insurance companies might want to know if you are at high or low risk, which could affect the premiums you have to pay. You will probably be able to think of many more possible advantages and disadvantages of having this knowledge.

17 Your plan diagram should:
- be large — preferably larger than the diagram in the book
- be drawn with clear, clean lines
- not show any individual cells
- be made up of four concentric lines — a pair of lines quite close together on the outside to represent the epidermis and root hairs; a line much closer to the centre of your drawing representing the boundary between the cortex and the endodermis; another line close to the last one representing the inner edge of the endodermis; and then a cross-shaped structure representing the xylem and phloem.

18 Loading sucrose into the phloem sieve tube.

19 Movement of water up xylem is down a water potential gradient, from roots to leaves. This gradient is maintained by transpiration in the leaves, which lowers the water potential there. The gradient is always in the same direction.

Movement of sap in phloem is down a pressure gradient, from source to sink. High pressure is produced by active loading of sucrose into the phloem at a source, which causes water to follow by osmosis. Different parts of the plant can act as sources at different times, for example leaves when they photosynthesise and roots when the starch stored in them is broken down. Different parts can also act as sinks at different times, for example roots in autumn when they are building up starch stores and flowers when they are using sugars to produce nectar.

20 a Roots are sources; leaves and flowers are sinks.
b Leaves are sources; flowers and roots are sinks.
c Leaves are sources; fruits and roots are sinks.
d Roots are sources; leaves are sinks.

Index

Index